New Tools for Monitoring Genetic Diversity in Animals

Guest Editors

Giovanni Forcina
Qian Tang
M. Pilar Cabezas

MDPI

Basel • Beijing • Wuhan • Barcelona • Belgrade • Novi Sad • Cluj • Manchester

Guest Editors

Giovanni Forcina
Department of Life Sciences
University of Alcalá
Madrid
Spain

Qian Tang
Rowland Institute at Harvard
Cambridge
USA

M. Pilar Cabezas
Centre of Molecular and Environmental Biology (CBMA)
University of Minho
Braga
Portugal

Editorial Office

MDPI AG
Grosspeteranlage 5
4052 Basel, Switzerland

This is a reprint of the Special Issue, published open access by the journal *Animals* (ISSN 2076-2615), freely accessible at: https://www.mdpi.com/journal/animals/special_issues/New_Tools_for_Monitoring_Genetic_Diversity_in_Animals.

For citation purposes, cite each article independently as indicated on the article page online and as indicated below:

Lastname, A.A.; Lastname, B.B. Article Title. *Journal Name* **Year**, *Volume Number*, Page Range.

ISBN 978-3-7258-4805-8 (Hbk)
ISBN 978-3-7258-4806-5 (PDF)
https://doi.org/10.3390/books978-3-7258-4806-5

Cover image courtesy of Filippo Barbanera

Contents

About the Editors

Giovanni Forcina

Giovanni Forcina is a researcher based at the University of Alcalá (Spain) with broad interests in conservation and evolutionary biology. He is currently working on microbiomics and host–parasite interactions within the framework of the One-Health strategy.

Qian Tang

Qian Tang is a population geneticist based at the Rowland Institute at Harward with a solid expertise in molecular phylogeny and landscape genomics. He is particularly interested in exploring the association between evolution and environmental drivers. He is currently leading a project on moths.

M. Pilar Cabezas

M. Pilar Cabezas Rodríguez is an invited auxiliary researcher at the University of Minho (Portugal) who is specialized in marine biodiversity, systematics, and bioinformatics. Her research focuses on advancing the understanding of marine biodiversity through DNA-based tools, investigating the ecological and evolutionary processes that shape this diversity, including the impacts of cryptic species and biological invasions. She is currently involved in a Blue Bioeconomy Pact project monitoring benthic macroinvertebrates to derive the biotic indices for an ecological status assessment.

Preface

This Reprint offers an overview of the latest genetic and genomic tools used to assess and track genetic diversity over time in the animal kingdom. It aims to be of use to researchers from different areas, such as conservation biology and animal production, who wish to keep pace with the rapidly growing suite of increasingly powerful and scalable techniques to obtain genome-wide data from their studied species.

Giovanni Forcina, Qian Tang, and M. Pilar Cabezas
Guest Editors

animals

MDPI

Article

Assessing the Seasonal and Spatial Dynamics of Zooplankton through DNA Metabarcoding in a Temperate Estuary

Jorge Moutinho [1,2,*], Diego Carreira-Flores [1,2], Pedro T. Gomes [1,2], Filipe O. Costa [1,2] and Sofia Duarte [1,2,*]

[1] Centre of Molecular and Environmental Biology (CBMA) and ARNET—Aquatic Research Network, Department of Biology, University of Minho, Campus Gualtar, 4710-057 Braga, Portugal; diego.carreira@bio.uminho.pt (D.C.-F.); pagomes@bio.uminho.pt (P.T.G.); fcosta@bio.uminho.pt (F.O.C.)

[2] Institute of Science and Innovation for Bio-Sustainability (IB-S), University of Minho, 4710-057 Braga, Portugal

* Correspondence: jorge.er.moutinho@gmail.com (J.M.); sduarte@bio.uminho.pt (S.D.)

Simple Summary: The routine monitoring of zooplankton is difficult due to their small size and morphological ambiguity. Also, the eggs and larva of meroplankton resemble one another, and therefore it is challenging to identify their species. Alternatively, DNA-based tools can provide precise species identifications, regardless of the size or the developmental stage of a specimen. We developed a protocol for testing the potential of DNA metabarcoding for assessing the seasonal and spatial dynamics of zooplankton in a temperate estuary. Both the seasonal and spatial gradients influenced recovered richness, composition, and taxonomic distinctness, confirming the great aptitude of DNA metabarcoding for providing higher density monitoring and shedding new light on the composition and dynamics of complex zooplankton.

Abstract: Zooplankton are key components of estuarine trophic networks. However, routine monitoring is hindered by the difficulty of morphology-based identification. DNA-based methods allow us to circumvent some of these hurdles, providing precise species identifications regardless of the taxonomic expertise of the investigator or the developmental stage of the specimens. However, the process is dependent on the completeness of the reference libraries. In this study, we sought to evaluate the potential of DNA metabarcoding to assess the seasonal (summer, autumn, and early spring) and spatial dynamics of zooplankton (four locations spanning ca. 6 km) in the Lima estuary (NW Portugal). Two genetic markers were used: the cytochrome c oxidase subunit I and the V4 hypervariable region of the ribosomal 18S rRNA genes. Overall, 327 species were recovered, and both markers displayed minute overlap (7% were detected with both markers). Species richness, composition, and taxonomic distinctness were majorly influenced by the season, with a declining tendency from summer (highest number of exclusive species, n = 74) to spring. Second to season, the taxa composition was influenced by spatial variation where the most downstream site displayed the highest number of exclusive species, n = 53. A total of 16 non-indigenous species were detected using metabarcoding, but only one (*Austrominus modestus*) has been documented out in the estuary. In conclusion, both the seasonal and spatial gradients influenced the recovered richness, composition, and taxonomic distinctness, confirming the great aptitude of DNA metabarcoding for providing higher density monitoring and shedding new light on the composition and dynamics of complex zooplankton communities.

Keywords: DNA metabarcoding; cytochrome c oxidase subunit I; hypervariable region V4 18S; zooplankton; biomonitoring; coastal ecosystems; Lima River estuary; non-indigenous species

1. Introduction

Monitoring zooplankton diversity is crucial to assessing ecosystem health and the impacts of environmental changes. Most biomonitoring studies have documented the responses of macro-eukaryotes to environmental alterations [1–3], while mostly neglecting small zooplankton [4].

Zooplankton play a significant role in the dispersion and distribution of energy between lower trophic levels, such as bacteria and eukaryotic phytoplankton, and higher trophic levels (macrofauna); this is known as bottom-up control; thus, they play a key role in the carbon biogeochemical cycle of coastal ecosystems [5–9]. Because zooplankton are a key component within aquatic trophic networks, monitoring their community will provide important information including the composition of holoplankton and meroplankton (eggs and larvae of benthic and nektonic organisms). From that information, we can assess the trophic state of the zooplankton community as it undergoes seasonal changes. Ultimately, we will improve our understanding of how climate change impacts marine and coastal ecosystems [10–14]. For instance, these communities have been reported to be sensitive to changes in water levels of reservoirs, trophic changes, and water quality, where commonly used WFD metrics were found to be insufficient for comprehensive quality assessments in these locations [15]. Similarly, in several estuarine systems, zooplankton were found to quickly respond to environmental conditions and anthropogenic pressures [16–19]. In addition, meroplankton often represent a significant component of the zooplankton community; thus, they can influence the location and potential yield of pelagic fisheries, either as food or nurseries [20], which can be associated with estuaries and other transitional waters. Furthermore, an analysis of the meroplankton community may represent an indirect way to assess the benthic invertebrate and fish communities and their reproductive cycles [21], providing a unique dataset that is otherwise difficult to obtain.

Researchers have recognized that the zooplankton's taxonomic composition is important to achieving an understanding of the links between the physical environment and higher trophic levels [14]. However, the morphological ambiguity of the early developmental stages (i.e., eggs, larvae, and juvenile stages) and the small size of zooplankton are serious challenges to the identification of their species based on morphology alone. Thus, the high phylogenetic diversity inherent in zooplankton samples, coupled with the effort needed to identify the specimens' species, limits research. On the other hand, DNA metabarcoding and high-throughput sequencing [22,23] have revolutionized the way that researchers characterize biodiversity in different types of ecosystems [23]. Indeed, it can greatly improve zooplankton identification by discriminating between morphologically similar species, the morphological plasticity of certain taxa, and overcoming the ambiguity of the early development stages. Further, DNA metabarcoding is a powerful tool that can be used to respond quickly to the needs of environmental managers; indeed, it already has been realized as a tool for large-scale biodiversity analysis [24,25], including the analysis of zooplankton [26,27]. Yet certain pitfalls are still present in metabarcoding schema: e.g., the lack of identification of individual life stages and the relative abundances of species [25,28]. Nevertheless, metabarcoding can display a high discrimination of spatial and temporal patterns in metazoan planktonic assemblages [27,29,30], and it can resolve hidden diversity: e.g., rare, low-abundant, and newly introduced taxa, as well as hard-to-identify meroplankton [31–35]. More recently, the potential of using the metabarcoding approach to provide biomass estimates has been used with several taxonomic groups, albeit with different degrees of success [34,36–38]. Most metabarcoding studies on zooplankton biodiversity use a single molecular marker to recover diversity [31,32]; this limits taxonomic scope [39]. Still, several studies have employed multiple molecular markers and multiple primer sets, which have displayed their marker complementarity for screening zooplankton taxa [30,40,41].

Here we report a study that used multi-marker DNA metabarcoding to monitor zooplankton in the Lima River estuary, a coastal ecosystem located in northwest Portugal. Specifically, we assessed both the seasonal and spatial variations of zooplankton and identified non-indigenous species.

2. Materials and Methods

2.1. Description of the Study Area

The Lima River estuary is temperate and drains into the Atlantic Ocean in the vicinity of the city of Viana do Castelo, in the northwest of Portugal (41°40′ N and 8°50′ W). It is characterized by a semidiurnal and mesotidal regime (3.7 m), with an average flushing rate of 0.40 m s^{-1}, river flow of 70 m^3 s^{-1}, and hydraulic residence of nine days [16,42,43]. Due to the human presence on the shores, it has a quite consistent width of around 400 m; however, the upstream part is shallower and wider, reaching a maximum of 1 km in width [44]. The present study was conducted throughout the estuary and four sites were selected: LMZ1, LMZ2, LMZ3, and LMZ4 (Figure 1, Table S1). LMZ1 was the most upstream site. Located at the mouth of the river, LMZ4 was the most downstream. These two sites are ~6 km apart. In between the upper and lower reaches are two sites that have been highly modified by human intervention (LMZ3 and LMZ4). The lowermost part of the estuary encompasses several man-made structures and is subjected to the continuous dredging of the navigation channel [45]. Upstream from the Eiffel Bridge (close to LMZ3), the estuary has retained most of its natural banks, with shallow saltmarshes and tidal sandy islands (Figure 1). In general, the estuary receives industrial, agricultural, and urban discharges of nutrients and other materials [46].

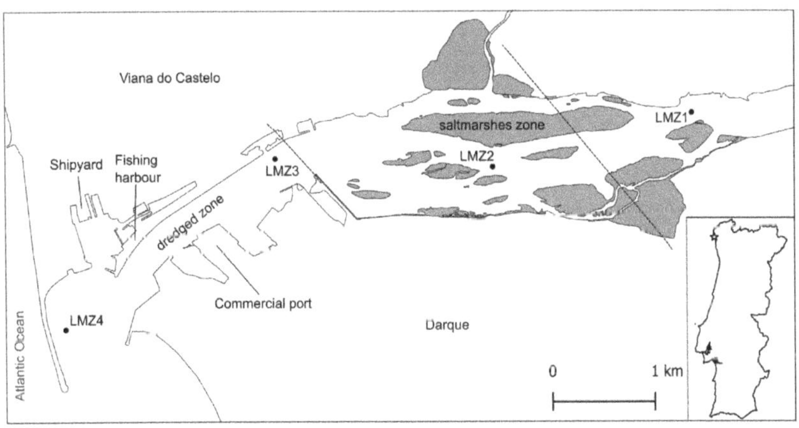

Figure 1. Sampling sites located in the Lima estuary, in the vicinity of Viana do Castelo, in the northwest of Portugal. On the map, saltmarshes are represented in grey and both bridges crossing the estuary are represented as dashed lines.

2.2. Sampling Strategy

Sample collection took place during high tide in two different points of the estuary characterized by more predominant human activity (LMZ4: river mouth and LMZ3: recreational marina) and in two additional points located upstream of the saltmarshes (LMZ1 and LMZ2), where dredging activity is null, which limits boat traffic (Figure 1). Zooplankton was sampled using oblique tows and a standard plankton net with a 50 cm opening diameter, 150 cm length, and a mesh size of 153 µm. At each sampling location, three separate tows were performed, after which the end-cup content was poured into a storage bottle, previously washed with bleach (10%) and MilliQ water. Any considerable residual content inside the end-cup was washed into the bottle with water from the respective site. This process was conducted across three different seasons: summer (27 July 2021), autumn (17 November 2021), and spring (24 May 2022), resulting in a total of 36 zooplankton samples. The salinity, conductivity, and pH parameters were measured with a WTW Multiline F/set 3 no. 400327 (WTW, Weilheim, Germany) from superficial water samples (Table S1, Supplementary Material). The wave and wind conditions did not provide the right conditions to measure the water parameters from the river mouth during summer sampling.

All measured physical parameters displayed a decreasing tendency during the summer–spring period and throughout all sampled locations, although there were a few exceptions for LMZ3 and LMZ4. At the most downstream location (LMZ4), the salinity, conductivity, and pH increased considerably from autumn to spring. The highest values were attained at LMZ2 during the summer and autumn for all parameters, while in spring, the maximum values were found at LMZ4, the most downstream location (Table S1, Supplementary Material).

The zooplankton samples were stored in sterilized plastic flasks with water from their respective sampling site. The flasks were kept in a large container filled with ice until they reached the lab (2–3 h). When returned to the lab, we processed the samples immediately. Filtration was done through Merck-Millipore membranes (47 mm diameter, pore size of 45 µm) using an EZ-Fit™ Manifold filtration ramp equipped with three Microfil® funnels attached to an EZ-Stream Vacuum Filtration Pump (Merck-Millipore, Darmstadt, Germany). All removable parts were submerged in bleach (10%) and rinsed with MilliQ water, and all surfaces of the working station were cleaned with bleach (10%) and ethanol (96%). To limit additional possibilities of cross-contamination between samples, we flamed the porous stone of the funnel base, which had previously been immersed in 96% ethanol. The filter membranes were then preserved at −20 °C until DNA extraction.

2.3. DNA Extraction, PCR Amplification, and High-Throughput Sequencing (HTS)

DNA was extracted from filtered samples using the DNeasy PowerSoil Kit, from Qiagen (Hilden, Germany), following the manufacturer's protocol with minor changes. Two technical replicates were considered for each sample, consisting of $\frac{1}{4}$ of the scraped-off zooplankton from the filters. After extraction, and for each sample, the two technical replicates were pooled together (30 µL of each). Negative controls were introduced during this step using exactly the same procedure but with new filters for checking for any contamination of the solutions of the DNA extraction kits and labware materials used. These negative controls were further used during PCR reactions. DNA concentrations were quantified using a NanodropTM 1000 spectrophotometer (Thermo Fisher Scientific, Waltham, MA, USA) (Table S2, Supplementary Material) and stored at −20 °C until PCR amplification and high-throughput sequencing.

Samples were prepared for Illumina MiSeq sequencing by targeting the eukaryotic communities through the amplification of the 18S rRNA and COI genes at Genoinseq (Biocant, Cantanhede, Portugal). Two different sets of primers were used: the forward primer mICOIintF 5′-GGWACWGGWTGAACWGTWTAYCCYCC-3′ [47] with the reverse primer LoboR1 5′-TAAACYTCWGGRTGWCCRAARAAYCA-3′ [48], which targets the 3′ region of COI (~313 bp); and the forward primer TAReuk454FWD1 5′-CCAGCASCYGCGGTAATTCC-3′ with the reverse primer TAReukREV3 5′-CTTTCGTTCTTGATYRA-3′ [49], which targets the hypervariable region V4 (~380 bp) of the eukaryotic 18S rRNA gene. Both primers were selected based on previous marine invertebrate analysis [50,51], where the use of a 3′ COI fragment (hereafter mentioned as COI) provided more resolved and reliable species-level identifications, which were even greater and had higher success rates than the versatile primer sets traditionally used for DNA barcoding (i.e., LCO149/HCO2198) [47,48,50–52], whereas the hypervariable region V4 of the 18S rRNA gene (hereafter mentioned as 18S) provided a broader scope on the recovered biodiversity. Still, the 18S species-level identifications should be considered with caution, since species-level resolution can be poor in some groups [53,54]. Although the herein-employed primers have been rarely mentioned in metabarcoding-based assessments of zooplankton diversity, both markers have been widely employed in metabarcoding-based zooplankton assessments [47,48,51,55].

For both markers, PCR reactions were performed for each sample using a KAPA HiFi HotStart PCR Kit according to manufacturer suggestions, 0.3 µM of each PCR primer, and 5 µL (COI) and 2.5 µL of template DNA in a total volume of 25 µL. For the COI, PCR conditions involved a 3 min denaturation at 95 °C, followed by 35 cycles of 98 °C for 20 s, 60 °C for 30 s, and 72 °C for 30 s, as well as a final extension at 72 °C for 5 min. For the 18S, PCR conditions involved a 3 min denaturation at 95 °C, followed by 10 cycles of 98 °C for 20 s, 57 °C for 30 s, and 72 °C for 30 s, 25 cycles of 98 °C for 20 s, 47 °C for 30 s, and 72 °C for 30s, and a final extension at 72 °C for 5 min. Negative PCR controls were included for all amplification procedures. The negative control samples did not amplify for any primer pair.

The DNA was further reamplified in a limited-cycle PCR reaction to add sequencing adapters and dual indexes to both ends of the amplified target region according to the manufacturer's recommendations (Illumina, San Diego, CA, USA). PCR products were then one-step purified and normalized using the SequalPrep 96-well plate kit (ThermoFisher Scientific, Waltham, MA, USA) [56], pooled and 250 bp paired-end sequenced in the Illumina MiSeq® (50,000 sequencing depth) sequencer with the MiSeq reagent Kit v3 (600 cycles), according to manufacturer's instructions (Illumina, San Diego, CA, USA) at Genoinseq (Cantanhede, Portugal). For further information regarding the amplification and sequencing steps, see Info S1 (Supplementary Material).

2.4. Bioinformatic Processing and Taxonomic Assignment

Quality filtration was performed on Illumina reads (fastq files) using PRINSEQ v0.20.4 [57] to remove sequencing adapters, trim bases with an average quality lower than Q25 in a window of 5 bases, and remove reads with less than 100 bases for 18S and 150 bases for COI. This initial processing was performed at Genoinseq.

Prior to taxonomic assignment, the filtered forward and reverse reads provided by the sequencing facility were merged by overlapping paired-end reads in mothur (*make.contigs* function, default) [58,59]. The resulting reads were then processed in two pipelines from public databases: the COI reads were submitted to the mBrave—Multiplex Barcode Research and Visualization Environment (www.mbrave.net; [60]; accessed on 19 December 2022) and the 18S reads were submitted to SILVAngs (ngs.arb-silva.de/silvangs/; [61]; accessed on 6 December 2022). In mBrave, the COI reads were trimmed by length (maximum 313 bp) and those with a minimum quality value (QV) higher than 10 were kept, which allowed for a maximum of 25% nucleotides with <20 QV and a maximum of 25% nucleotides with <10 QV. Reads were then taxonomically assigned using a similarity threshold of 97% against all the system's reference libraries, the personally curated and Iberia Peninsula-specific reference libraries, as well as taxa-specific reference libraries (December 2022)—all from the BOLD database [62] (see Info S2 (Supplementary Material) for the list of the COI reference sequence libraries).

The 18S reads were processed by the amplicon analysis pipeline of the SILVA project (SILVAngs 1.4) [61]. Each read was aligned using the SILVA Incremental Aligner (SINA v1.2.10 for ARB SVN (revision 21008)) [63], against the SILVA SSU rRNA SEED, and quality controlled [61]. Reads shorter than 200 aligned nucleotides and reads with more than 1% ambiguities or 2% homopolymers, respectively, were excluded from further processing. Putative contaminations and artifacts, and reads with a low alignment quality (80 alignment identity, 40 alignment score reported by SINA) were identified and excluded from downstream analysis. After these initial steps of quality control, identical reads were identified (dereplication), the unique reads (100%) were clustered (OTUs) on a per-sample basis, and the reference read of each OTU was classified. Dereplication and clustering were done using VSEARCH (version 2.17.0; https://github.com/torognes/vsearch; accessed on 6 December 2022) [64], applying identity criteria of 1.00 and 0.99, respectively. The classification was performed by BLASTn (2.11.0+; http://blast.ncbi.nlm.nih.gov/Blast.cgi; accessed on 6 December 2022) [65], with standard settings using the non-redundant version of the SILVA SSU Ref dataset as classification reference (release 138.1; http://www.arb-silva.de; accessed on 6 December 2022). The classification of each OTU reference read was mapped onto all reads that were assigned to the respective OTU using a 99% similarity threshold (December 2022). For further detailed information regarding both bioinformatic pipelines (mBrave and SILVAngs), see Info S1 (Supplementary Material).

Any read assigned to a non-metazoan taxon was discarded. The nomenclature of detected taxa was confirmed using the World Register of Marine Species database (WoRMS; www.marinespecies.org, accessed on January of 2023). Additionally, all BINs (Barcode Index Numbers) to which COI reads were assigned were thoroughly curated (identification of misidentifications and synonymized species names associated with the same BIN, as well as ambiguous groupings) to attain the most reliable identifications, particularly at the species-level. Regarding the 18S dataset, sequences were blasted against the NCBI's database to assess the reliability of the taxonomic assignment provided by the SILVAngs curated database [66]. Throughout further analysis and discussion, only species-level identifications were considered, due to the taxonomic uncertainty that can be associated with OTUs, while displaying more than 8 reads on each dataset [50,67].

2.5. Data Processing and Analysis

All statistics and graphics were performed with the Paleontological Statistics software (PAST, v4.09) and in the R environment (version 4.1.2.) using the Vegan package (version 2.6.4) [68], unless otherwise stated. Only presence/absence data was considered due to the putative amplification associated bias. Treemap was performed using the function *treemap* from the package of the same name [69] in order to present and determine the most relevant taxonomic groups recovered from the sampled zooplankton and further analyze the complementary effect of the multi-marker approach herein employed.

Beyond this point, we only considered the merged COI and 18S datasets due to the complementarity of both markers in their species detection. Both taxonomic diversity and distinctness were determined using the PAST software, which determines both taxonomic metrics based on Clarke & Warwick [70]'s definition and the author's own 1000 random replicates test, and can be represented by the following formula:

$$\Delta = \frac{\sum\sum_{i<j} w_{ij} x_i x_j}{\sum\sum_{i<j} x_i x_j},$$

where w_{ij} are weights varying if concurrent species (0) or different species (1), and x are the abundances. Due to the nature of this study, both the taxonomic diversity and distinctness were represented with the same values (hereafter represented solely as taxonomic distinctness). Species richness and taxonomic distinctness fluctuations were assessed using a 2-way ANOVA, followed by post-hoc (Tukey's HSD) analysis to assess the differences among the levels of each factor (season and location) using the *aov* and *TukeyHSD* functions, respectively. Non-metrical Multidimensional Scaling (nMDS) ordination was used to examine recovered community composition similarities among samples using Jaccard's dissimilarity index. The influence of spatial/seasonal sampling was tested using the *adonis2* function (2-way PERMANOVA; 999 permutations). Prior to all multivariable analysis, the *decostand* function was used to transform read datasets into presence/absence, since the *metaMDS* (function used to perform nMDS) determines Jaccard's dissimilarity based on Bray-Curtis [68]. The *cor* function was used to determine Pearson's correlation of each species recovered with the dimensions used for the nMDS (n = 2), which was then used to determine the taxa that better fit the attained clustering, based on p value, using the *envfit* function. Furthermore, a presence–absence heatmap and clustering was performed—with the number of samples in which each taxon was recovered—using the *pheatmap* function of an R package of the same name [71], in order to assess the species-level clustering throughout the spatial and seasonal gradient. All previous analyses were repeated to analyze the most relevant taxonomic groups determined by the previous treemap analysis.

All Venn diagrams, used for the visual representation of the partitioning of the recovered zooplankton diversity, were developed using the InteractiVenn platform [72].

3. Results

3.1. HTS Data Initial Processing and the General Taxonomic Composition Recovered by Each Marker

The high-throughput sequencing of the 36 samples resulted in a total of 2,096,992 and 1,548,889 recovered COI and 18S reads, respectively, from which, after filtration and quality checking, around 72% of COI reads were eligible for taxonomic assignment, but for 18S, less than half of the reads were eligible for taxonomic assignment (42%) (Table S3, Supplementary Material). Indeed, this was particularly observed in samples from the spring season collected at the most downstream location (LMZ4).

Around 36.4 and 21.2% of COI and 18S reads, respectively, were taxonomically assigned to the aquatic metazoan. However, for a large portion of these, obtaining species-level assignments was not possible, resulting in their classification at higher taxonomic ranks. A total of 15 metazoan phyla were recovered (Figure 2A), from which Hemichordata was specifically recovered with COI, while Entoprocta, Chaetognatha, Nematoda, Platyhelminthes, and Phoronida were 18S-exclusive records (Figure 2B). Consequently, 18S dominated in the number of exclusive detections throughout the whole taxonomic spectrum (except for class and species ranks), particularly at the order level (almost two times greater than COI; Figure 2A). The 18S:COI ratio of exclusive taxonomic assignments was found to be lower the greater the taxonomic resolution obtained. On the other hand, markers' complementarity was shown to be greater the higher the resolved taxonomic assignment (Figure 2A). Indeed, only 466,724 of the COI reads (ca. 22.3%) and 306,648 of the 18S reads (ca. 19.8%) were taxonomically assigned to species level, resulting in the detection of 175 species each, from which only 23 species (7%) were found to have been recovered with both markers (Figure 2A; for further details see Table S4, Supplementary Material), which represented less than half of the overlap observed at the genus level (Figure 2A).

From the total of 327 recovered species, 85% belonged to five different phyla: 49 and 55 species to Mollusca, 44 and 42 species to Arthropoda, 27 and 29 species to Annelida, 18 and 13 to Cnidaria, and 11 and 14 species to Chordata (for COI and 18S, respectively) (Figure 2B). The remaining recovered taxa represented 10 phyla, namely Bryozoa, Echinodermata, Nematoda, Nemertea, and Porifera, recovered with both markers, as well as Chaetognatha, Platyhelminthes, Phoronida, and Entoprocta 18S-exclusive, and Hemichordata COI-exclusive. The latter three phyla were solely represented by a singular species each. Arthropoda diversity was dominated by Copepoda (39%), followed by Decapoda (23%) and Thecostraca (15%). The 18S displayed greater taxonomic coverage of Copepoda than COI—18S recovered six different Copepoda orders (three exclusive: Monstrilloida, Polyartha, and Siphonostomatoida), as opposed to the three recovered with COI (Calanoida, Cyclopoida, and Harpacticoida). Both markers displayed greater representations of Calanoida during the study (61% and 69% of the Copepoda, for 18S and COI), followed by Harpacticoida (two 18S and three COI exclusive species) and Cyclopoida (two 18S and a COI exclusive species). Gastropoda (74% with COI and 34% with 18S) and Polychaeta (similar recovered richness with both markers) encompassed around 71.2% and 94.3% of the Mollusca and Annelida, respectively. Ascidiacea represented a fourth of the herein Chordata recovered, but most were detected by 18S. The remaining 10 phyla encompassed no more than a total of 10 species each (12.2%).

Because the two markers were complementary in species and taxonomic group detection, we opted to conduct the remaining analyses using the species-level dataset where data recovered with both markers were joined together, which allowed us to see more clearly the spatial and seasonal patterns of zooplankton in the Lima estuary.

Overall, around 32.4% of the recovered species consisted of single-sample recoveries, which included a greater representation of the overall Chordata and Porifera (72% and 60%), as well as half of the recovered Platyhelminthes, Nematoda, and Nemertea species. The only recovered Hemichordata species, *Balanoglossus clavigerus* (Delle Chiaje, 1829), was also detected in a single sample. On the other hand, only 28 species (ca. 8.6%) were recovered in more than 50% of the samples, with the majority being composed of

Arthropoda (13 species) and Mollusca (9 species); however, only 13 species were recovered in the majority of samples (herein 75%), which accounted for nine Arthropoda species: the Calanoida *Temora longicornis* (Müller O.F., 1785) and *Pseudocalanus elongatus* (Brady, 1865), and the Thecostraca *Austrobalanus imperator* (Darwin, 1854), *Verruca stroemia* (O.F. Müller, 1776), *Austrominius modestus* (Darwin, 1854), *Chthamalus montagui* (Southward, 1976), *Amphibalanus improvisus* (Darwin, 1854), *Balanus trigonus* (Darwin, 1854), and *Sacculina carcini* (Thompson, 1836); three Mollusca species: the Gastropoda *Peringia ulvae* (Pennant, 1777), the Bivalvia *Hiatella arctica* (Linnaeus, 1767) and *Mytilus* sp., and the Hydrozoa *Obelia dichotoma* (Linnaeus, 1758).

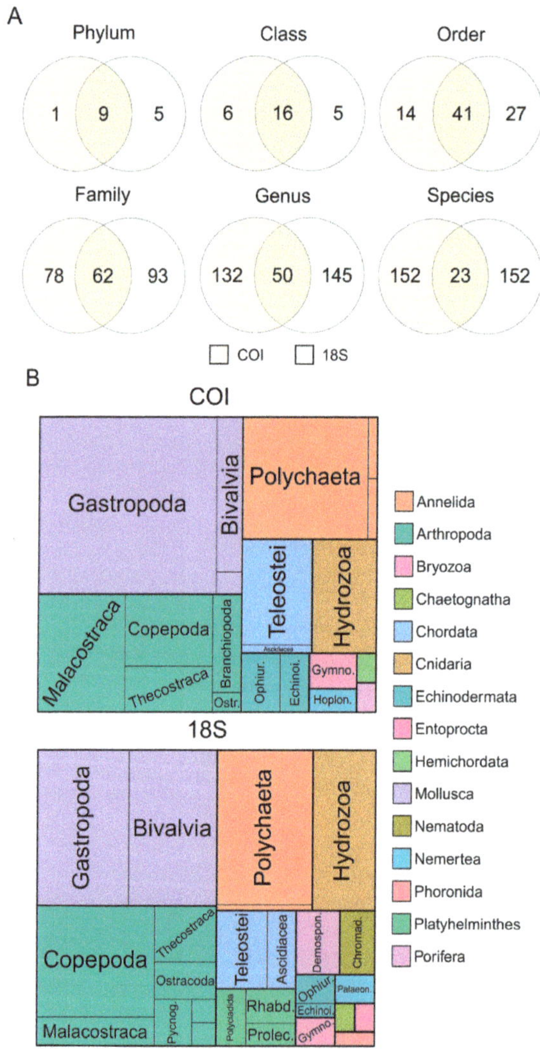

Figure 2. Partitioning of recovered taxa with COI and 18S (>97% and >99% similarity, respectively), depicted in all major taxonomic ranks (phylum, class, order, family, genus, and species) (**A**). Zooplankton composition, recovered through HTS using COI and 18S primers (**B**). The latter is color-coded based on recovered phyla. For further details, see the list of recovered taxa found in Table S4 (Supplementary Material).

3.2. Non-Indigenous Species Detection

DNA metabarcoding allowed the detection of 16 NIS, which accounted for around 4.9% of the recovered zooplankton biodiversity (Table S4, Supplementary Material). The bulk of recovered NIS were Arthropoda (7 species—*Amphibalanus eburneus* (Gould, 1841), *A. improvisus*, *A. modestus*, *B. trigonus*, *Eriocheir sinensis* (H. Milne Edwards, 1853), *Oithona davisae* (Ferrari F. D. & Orsi, 1984), and *Pseudodiaptomus marinus* (Sato, 1913)), followed by Ascidiacea (*Ciona intestinalis* (Linnaeus, 1767), *Microcosmus squamiger* (Michaelsen, 1927) and *Styela plicata* (Lesueur, 1823)), Mollusca (the gastropod *Crepidula fornicata* (Linnaeus, 1758), and the bivalves *Mercenaria mercenaria* (Linnaeus, 1758) and *Ruditapes philippinarum* (A. Adams & Reeve, 1850)). The remaining NIS included the polychaete *Pseudopolydora paucibranchiata* (Okuda, 1937), the bryozoan *Tricellaria inopinata* (d'Hondt & Occhipinti Ambrogi, 1985), and the hydrozoan *Cordylophora caspia* (Pallas, 1771). Overall, every recovered NIS consisted of exclusive detections by either the COI or the 18S (8 and 6 exclusive NIS recovered, respectively), except the copepod *P. marinus*, which was detected with both markers. For further details, see Table S4 (Supplementary Material).

3.3. Seasonal and Spatial Dynamic Effects on Species Richness, Taxonomic Composition, and Distinctness

Overall, average species richness displayed a decreasing tendency through all the study periods (ANOVA, F = 27.9, $p < 0.01$), but, at LMZ2 and LMZ3, species richness displayed a different response to seasonal variation compared to that of LMZ1 and LMZ4, which declined through all seasons (Figure 3A). The trending decline in species richness attained was indeed observed at the intermediate sites: at LMZ2, a strong decline in species richness was observed from summer to autumn, but not from autumn to spring; while at LMZ3, the opposite pattern was observed: species richness slightly increased from summer to autumn but decreased in spring. Spatially, the recovered species richness showed an overall increment toward downstream peaking at LMZ3, followed by a steep decline at LMZ4, supported by Tukey's post-hoc ($p < 0.05$, comparison between LMZ3 and LMZ4), which was a pattern observed more particularly in autumn and spring samples (Figure 3A).

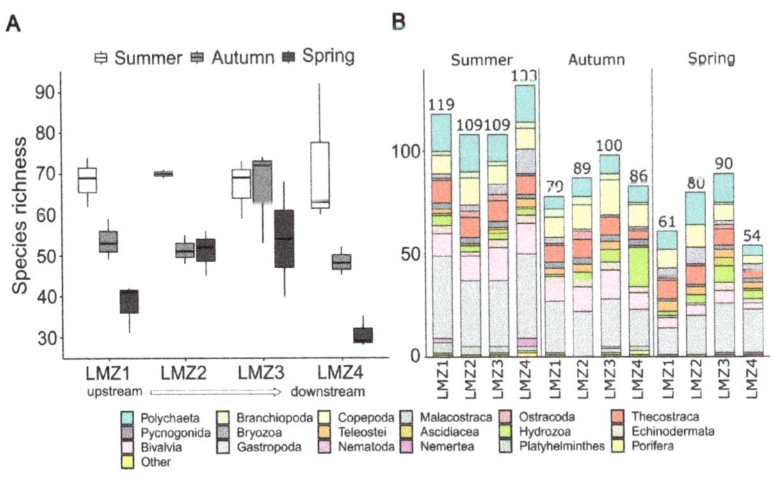

Figure 3. Seasonal and spatial influences on recovered zooplankton species richness (**A**) and taxonomic composition of the species recovered at all sites throughout the study duration (**B**). N = 3. In (**B**), data obtained in all replicates were joined together before analysis.

Furthermore, the effect of seasonal and spatial variation diverged between different taxonomic groups, particularly those encompassing meroplankton. For instance, Polychaeta (Annelida) were more well represented during the summer and spring (in the latter at LMZ2 and LMZ3), while Mollusca were better represented in the summer, in particular Gastropoda (Figure 3B). Hydrozoa (Cnidaria) were recovered throughout the study duration; however, a greater number of species was observed in the LMZ4-autumn, which was characterized by the recovery of 19 species, of which 13 were exclusive to these samples (coincided with the 3 exclusive Pycnogonida records).

On the other hand, the taxonomic distinctness demonstrated, in general, a different pattern from that observed for species richness. For instance, throughout the spatial gradient of the study area, a trending increase in taxonomic distinctness was observed from the upstream to downstream sites during all three sampled seasons (ANOVA, F = 3.16, $p < 0.05$) (Figure 4A), although no influence was found from season. Indeed, at every site, the taxonomic distinctness was maintained stably throughout the study duration. Nevertheless, the taxonomic distinctness was, in general, fairly stable, ranging from 5.39 (at LMZ1-summer) to 5.63 (at LMZ3-spring).

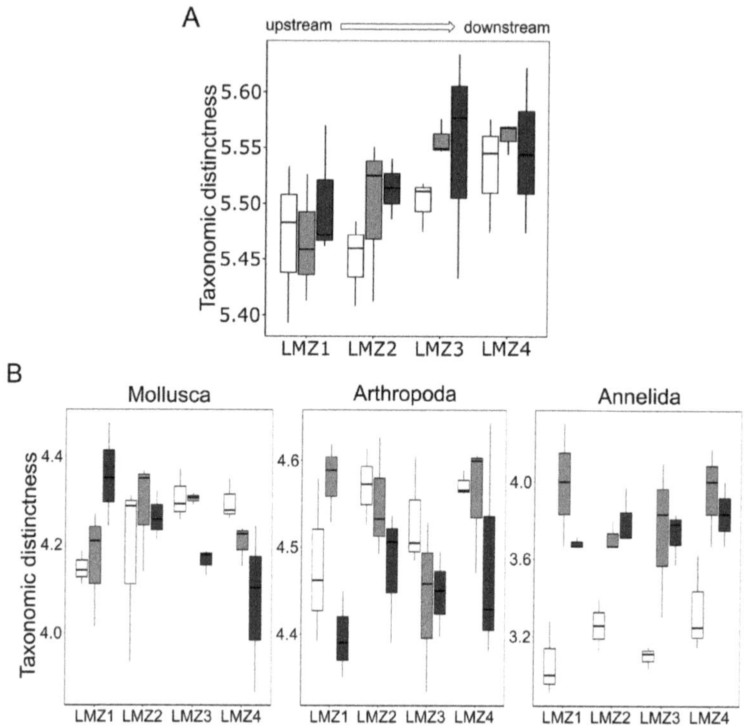

Figure 4. Seasonal and spatial influences on recovered zooplankton taxonomic distinctness (**A**), and group-specific taxonomic distinctness of the most relevant phyla based on species richness (Mollusca, Arthropoda and Annelida) (**B**).

The taxa-specific taxonomic distinctness also demonstrated divergent patterns compared to the general analysis of zooplankton diversity (Figure 4B). Annelida and Arthropoda taxonomic distinctness was demonstrated to be more influenced by seasonal variation (ANOVA, F = 32.98, $p < 0.01$; F = 4.95, $p < 0.05$, respectively), but not by the estuary's spatial gradient. However, both the Annelida and Arthropoda demonstrated diverging patterns from each other: the former displayed much lower values in the summer than in the remaining seasons, while the latter revealed a decreasing tendency throughout the study's duration (from summer, autumn, and spring) (Figure 4B). Moreover, the Mollusca taxonomic distinctness revealed an opposite trend compared to that of the general zooplankton diversity, decreasing from LMZ1 toward LMZ4, but no significant differences were found in such variation nor from a seasonal effect (ANOVA, $F = 0.18, p > 0.05$; $F = 0.64, p > 0.05$, respectively). The taxonomic distinctness was indeed highly dependent on the taxonomic group analyzed, and varied differently for different taxonomic groups, e.g., for Annelida, the minimum and maximum taxonomic distinctness scored higher than 1, while for Arthropoda and Mollusca such an interval was much lower (Figure 4B).

The partitioning of the species across all seasons (in general and for each site, separately, Figure 5) and across all sites (in all seasons and for each season separately, Figure 6) is displayed in Venn diagrams. In general, only 53 species were recovered in all seasons and the highest number of exclusive species was detected in summer (78 species). The highest number of species was shared between summer and autumn (41 species), while the lowest number was shared between autumn and spring (13 species), and this pattern was observed throughout the estuary, at all sampled sites (Figure 5). In addition, the highest number of exclusive species was detected in summer (66, 47, and 83 species for LMZ1, LMZ2, and LMZ4) at all analyzed sites, with the exception of LMZ3, where the highest number of exclusive species was detected in spring (44 species). On the other hand, 92 species were detected at all sampled estuarine sites (60 in summer, 33 in autumn, and 20 in spring). The highest number of exclusive species was recovered at LMZ4 (53 species, approx. 16%), while the lowest was at LMZ1 (21 species) (Figure 6). The highest number of shared species was found between LMZ3 and LMZ4 in the summer and autumn (9 and 8 species, respectively); while in spring, the highest number of species (9 species) was shared between the most upstream sites (LMZ1 and LMZ2) (Figure 6).

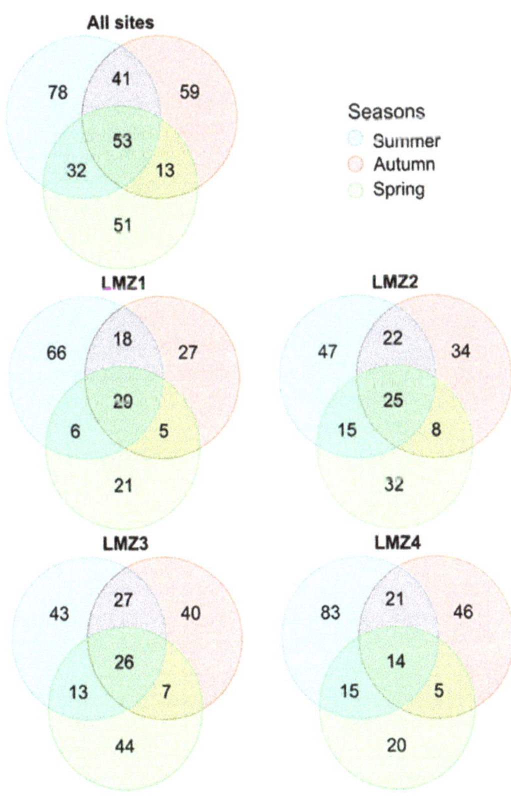

Figure 5. Seasonal partitioning of zooplankton diversity recovered with metabarcoding throughout the whole sampled spatial extension and for each sampling site: LMZ1, LMZ2, LMZ3, and LMZ4.

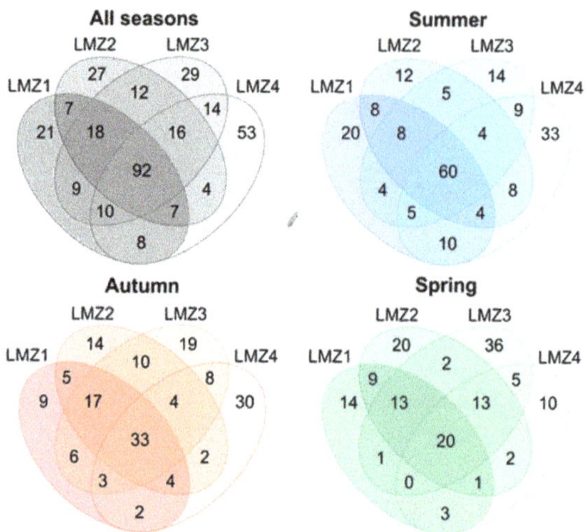

Figure 6. Spatial partitioning of zooplankton species recovered with metabarcoding throughout the study duration and for summer, autumn, and spring samples. From the most upstream site (LMZ1) to the most downstream (LMZ4).

3.4. Seasonal and Spatial Dynamics of Zooplankton Structure Composition

A non-metric multidimensional scaling analysis indicated that the zooplankton recovered from the Lima estuary were mostly structured by season, but considerable spatial turnover was also observed (Figure 7 and Table 1). Both the summer and spring samples displayed closer relations to the autumn cluster (NMDS2 and NMDS1, respectively) than with each other (pairwise PERMANOVA, $p < 0.01$), but still maintain their own characteristic compositions (pairwise PERMANOVA, $p < 0.01$ for both). LMZ4 was demonstrated to be the most species-level composition divergent site from the estuary, forming distinct clusters in all seasons (Figure 7), except in spring, where LMZ4 clustered with LMZ3 samples, thereby dividing the estuary into two different clusters (Figure 7). Still, in autumn, the most downstream location displayed the most divergent zooplankton composition from the study. Indeed, a high number of exclusive Hydrozoa and Pycnogonida were recovered, while Annelida contribution was lower in these samples (Figure 3B); however, the recovery of the Hydrozoa *Clytia gracilis* (Sars, 1851), *Clytia paulensis* (Vanhöffen, 1910), *Obelia longissima* (Pallas, 1766) and *Orthopyxis integra* (MacGillivray, 1842), the Pycnogonida *Achelia echinata* (Hodge, 1864), the Gastropoda *Trinchesia caerulea* (Montagu, 1804), the Annelida *Malacoceros fuliginosus* (Claparède, 1868), and the Chaetognatha *Parasagitta friderici* (Ritter-Záhony, 1911) were found to be significantly correlated to this ordination ($p < 0.01$) and better explained such a distribution (Table S5, Supplementary Material).

Furthermore, the species-level composition throughout the study differed between the taxonomic groups, with the exception of the Mollusca, for which the observed seasonal and spatial patterns were, indeed, very similar to those found for general zooplankton (Figure 7 and Table 1). Although summer and autumn samples displayed their own clustering (NMDS2), both were more related to each other than to the spring Mollusca zooplankton composition (NMDS1), though Mollusca composition along the estuary spatial gradient varied less. In both the autumn and spring clusters, the LMZ4 samples were closer to those from the remaining sites.

On the other hand, zooplankton's Arthropoda and Annelida composition revealed differing patterns compared to the general zooplankton and Mollusca compositions (Figure 7). Arthropoda displayed distinct seasonal clustering, where summer and spring were highly similar to each other; however, a particular spring cluster was formed for NMDS1, encompassing the LMZ1-LMZ3 samples, and separating autumn from summer.

An opposite pattern was observed for Annelida, where the autumn and spring clusters were more closely related to one another than to summer samples, although to a lesser extent. The LMZ4 species-level composition formed distinct clusters for both the Arthropoda and the Annelida, although within the latter a greater variation was displayed (Figure 7).

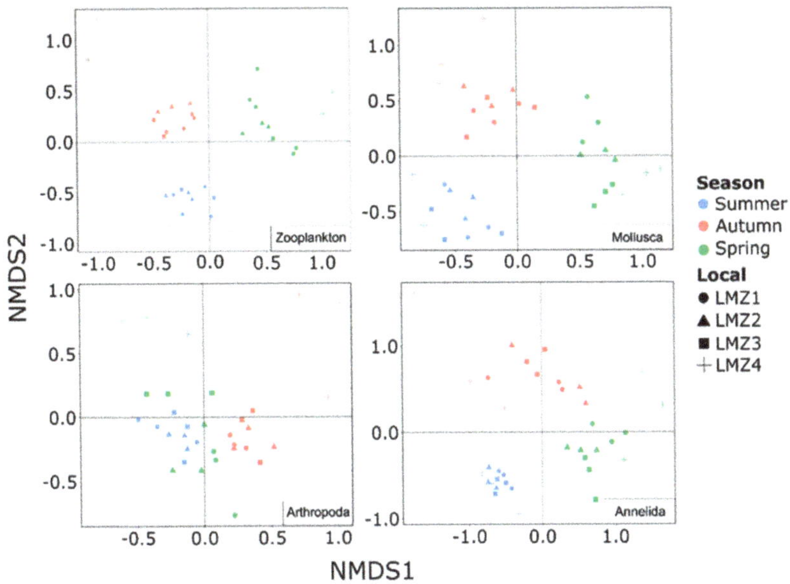

Figure 7. nMDS ordination of Lima estuary's zooplankton recovered through metabarcoding employing a multi-marker approach (A; Stress = 0.13), and the most relevant phyla (Mollusca: Stress = 0.16; Arthropoda: Stress = 0.17; Annelida: Stress = 0.14). All similarity matrices used were based on presence–absence data (Jaccard's dissimilarity index).

Table 1. Permutational analysis of variance (PERMANOVA) results of the most relevant taxonomic groups from the recovered zooplankton with DNA metabarcoding (999 permutations, based on Jaccard's dissimilarity index).

Taxonomic Group	Season		Site	
	F	*p*	F	*p*
Zooplankton	9.57	<0.01	2.54	<0.01
Mollusca	10.91	<0.01	1.91	<0.01
Arthropoda	7.26	<0.01	3.97	<0.01
Annelida	11.18	<0.01	2.08	<0.01

Considering the relative abundance (number of samples recovered), species-level clustering generated consistent results with the nMDS for the whole dataset. The set of 327 zooplankton species revealed clustering in seven groups (G1-7), ranging from 6 to 233 species, that appeared to better explain the stronger seasonal variation over spatial influence on the Lima estuary's zooplankton (Figure 8). Indeed, the G4 (27 species), G5 (14 species), and G7 (20 species) clustered species with greater representation in summer, autumn, and spring, respectively—although a G4-subgroup included summer/autumn related species—and G3 (8 species) better represented summer/autumn, while species found throughout the whole seasonal and spatial gradient, with a remarkable representation in all samples, were clustered within G1 (6 species) and G2 (19 species). G6 was the broadest cluster, which included 233 species, where sample-specific species and species with lower representation were clustered together.

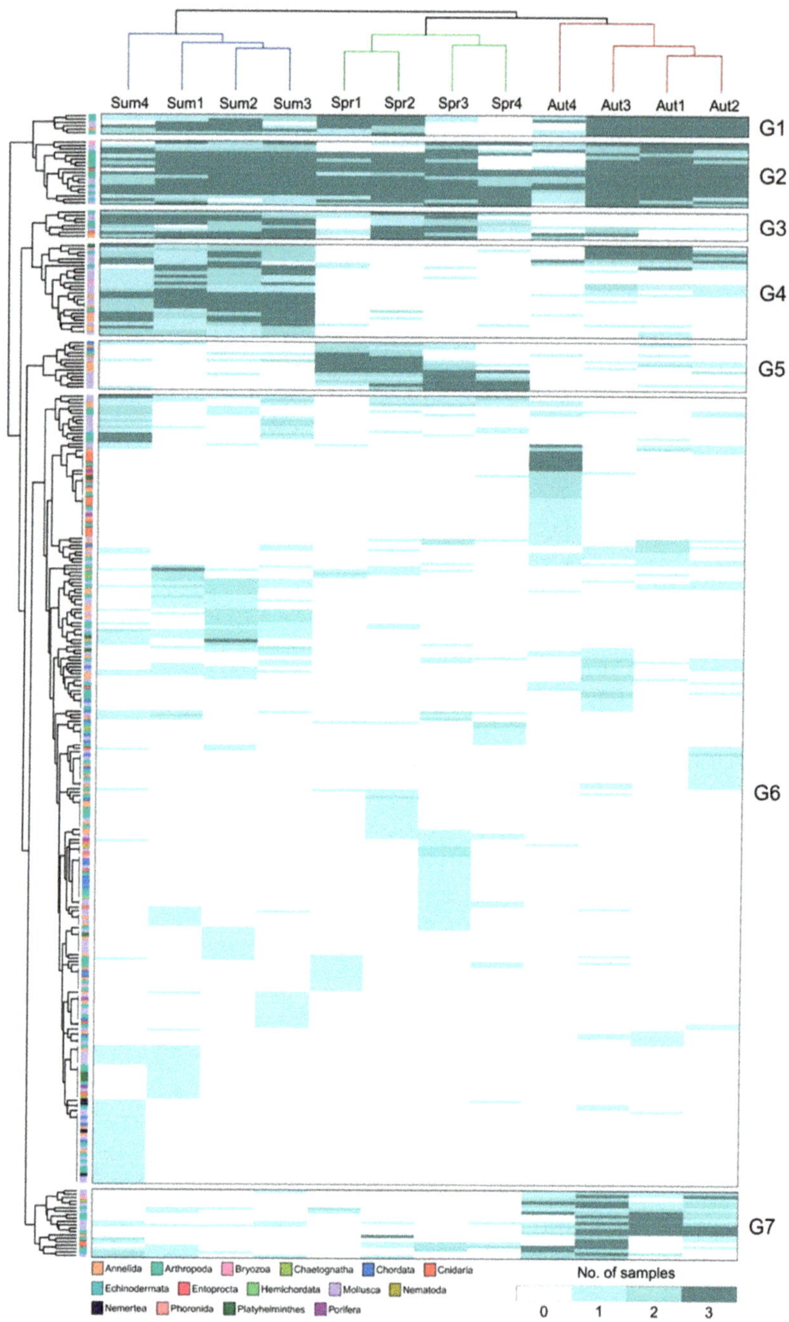

Figure 8. Species-level (y-axis) and sampling event (x-axis) clustering associated with a heatmap color-coded based on the number of replicates where each taxon was detected on each sampling event (0–3, with 0 indicating absence in all replicates and 3 presence in all replicates). Both dendrograms are color-coded based on each taxa phylum for the y-axis and based on the sampled seasons for the x-axis (blue = summer, green = spring, red = autumn).

Spatial variation also had an effect on species clustering. For instance, the two-part division of the estuary observed in the spring was particularly found to be explained by G5 sub-groups and G1, while G6 sub-groups may have also had a small influence. In the summer, LMZ4 was displayed as the outgroup (similarly seen in Figure 7), but not as far as observed in the autumn samples. Such was not shown to be associated with species shift, but in fact with variations in the relative abundance, more particularly in the G1 and G4 clusters. The high dissimilarity of LMZ4 in the autumn from the remaining sites above-mentioned further supported the influence of the exclusive recovery of several Hydrozoa and the Pycnogonida composing the G6 subgroup, as well as the loss and lower representation of several taxa throughout G1, G2, G4, and G7. For further details regarding clustering of species, see Table S6 and Figure S1 (Supplementary Material).

4. Discussion

Our study had three main outcomes: (i) We confirm the great utility of DNA metabarcoding for assessing the seasonal and spatial dynamics of zooplankton in a temperate estuary. (ii) The multi-marker approach allowed for a broader characterization of zooplankton diversity and allowed for the species-level identification of taxa that would otherwise be assigned to higher taxonomic ranks. (iii) We detected several NIS for the first time in the Lima estuary, demonstrating the utility of zooplankton metabarcoding for the early detection of NIS.

4.1. DNA Metabarcoding Performance in the Assessment of Zooplankton Species in the Lima Estuary

For conducting our analyses, we opted to use species-level assignments, which are crucial for mapping species occurrence and distribution and to compare to what has been found so far in the estuary. Although we are highly aware of the issues pertaining the completeness and reliability of current DNA barcode reference libraries [73], that can have a strong impact on the taxonomic characterization of using species level assignments on DNA metabarcoding, we point out several reasons for our choice: (i) a strong effort has been conducted in filling DNA barcode reference libraries for the studied region [74–77], in developing informatic tools for auditing COI reference records that were used in the current study (BAGS; see [78]), and in curating reference libraries [76,79,80]; (ii) gap-analyses of compiled lists of historic zooplankton and macrozoobenthos species in the study indicated very acceptable COI and 18S coverage in both BOLD and GenBank (97%, for zooplankton species and 85.2% for macrozoobenthos) [81]; and (iii) the reliability of species-level taxonomic assignments has been a posteriori verified based on currently available data (66% of the detected species are considered reliable).

Zooplankton assessments in the Lima estuary have been scarce and generally lacking in thorough and species-level taxonomic characterization. In fact, to the best of our knowledge, only one study had previously characterized the zooplankton community's structure in the Lima estuary, but with an emphasis on Copepoda diversity [16], while the remaining studies focused on either ichthyoplankton [42,82–84] or functional holoplankton and meroplankton abundance and biomass [85–87]. Thus, meroplankton characterization has been typically less resolved and overlooked through morphology-based analysis. Herein, metabarcoding-derived data revealed a greater diversity of planktonic metazoans occurring in the Lima estuary than what has been previously described from morphology-based historical reports [16,42,82,83,85–87]. A higher performance in biodiversity recovery was expected with DNA metabarcoding since its potential for uncovering unreported biodiversity, including that of meroplankton screening efficacy, has been shown in several other studies [31,33,88]. In addition, the recovered zooplankton from the Lima estuary have indeed improved the species-level identifications of meroplankton, such as Bivalvia, Gastropoda, Polychaeta, and Hydrozoa (Figure 2B), compared to previous morphology-based surveys, and further allowed the detection of additional occurring planktonic forms of taxa not yet reported in the estuary, such as Porifera [16,85] (Table S7, Supplementary Material).

However, the discordance between the herein-recovered taxa and historical records is indeed considerable. For instance, sampling effort (spatially and temporally) is one of the pivotal considerations taken into account in zooplankton characterization and it is likely that it may have influenced the resulting characterization—both spatial and temporal profiles of the current study did not fully overlap those from previous studies (i.e., the most recent study was six to seven years apart from sample collection in the current study). Species list compilation may also have influenced results, as only published studies were considered, whereas no input was considered from other sources, such as the Global Biodiversity Information Facility (GBIF) database, nor WoRMS. Furthermore, communities' fluctuations may have also played a role in the observed gap with species' introduction into/extinctions in the estuary. For instance, plankton periodically produce resting stages that sink and accumulate in the sediment. These diapausing stages are prevalent in estuaries and lagoons [89] and remain dormant for long periods of time before hatching. Thus, they may have played a role in the differences between results of our study and historical records. That is, the diversity of resting stages may be richer than the expressed one [90,91], including those of historical surveys and the current assessment through DNA metabarcoding. For instance, Rubino & Belmonte [91] reported 80 species in sediment samples, which were absent from plankton sampled from the water column.

Thus far, the uncovering of a high meroplankton diversity further supports the nursery role of the Lima estuary for several macrozoobenthos species. In fact, earlier zooplankton assessments demonstrated high meroplankton abundance/biomass in certain regions within the Lima estuary [85–87]. The hydrodynamics of the estuary may indeed play a larger role than what has been documented in previous assessments, despite the different sampling and identification methodologies employed [16,92]. In addition, the residence time and the semidiurnal and mesotidal regime of the Lima estuary generate highly suitable conditions for the development of a dominant stationary wave [84,93], which may allow enough time for the development of meroplankton inside the estuary [94]. Indeed, a higher abundance of several meroplankton groups in the Lima estuary has been reported than in the neighboring Minho estuary (average flushing rate of 300 m s^{-1}), [16], which may have accounted for the considerably high meroplankton diversity found at the Lima estuary through metabarcoding. Furthermore, some studies have been pointing out that the Lima estuary supports several fish species, relevant to local fisheries, as a nursery zone [42,84,95]. However, herein ichthyoplankton recovery resulted in 20 fish species from which some assignments might be questionable, particularly in the 18S dataset, since the 18S rRNA gene is too conserved across a broad range of fish species [22].

4.2. COI and 18S rRNA Gene Markers Displayed Minute Overlap in Zooplankton Species Detection

The taxa-overlap between both molecular markers was strong when considering the higher taxonomic ranks, but the more specific the identification the greater the complementarity found between both markers. Indeed, the proportion of exclusive detections with both markers was remarkably high, and only 23 species were taxonomically assigned in simultaneous COI and 18S reads. The high complementarity between the employed markers was expected, since previous studies have demonstrated a considerable range of exclusive detections for the species-level identification of marine zooplankton [30,41,52,88]. Generally, the mitochondrial COI has been used for more reliable species-level assignments of recovered reads and is the standard DNA barcode for the molecular-based identification of metazoans [96]—although at the cost of, e.g., the high prevalence of primers–templates mismatch leading to PCR associated bias [97,98]—while the nuclear 18S rRNA gene has been traditionally used in aquatic microbial eukaryote assessments [99,100], since it displays a broader taxonomic scope—hence, it has an extensive reference database. However, 18S sequences can be too conserved to discriminate organisms at low taxonomic levels, so it is usually used for broad-range high-rank taxonomic assignments [30,41,52,88]. In the present study, species-level assignments were not possible for 26,188 of the 18S reads; however, the COI allowed for greater resolution with species-level identifications of several 18S taxa that were only assigned to genus level, namely, *Balanus*, *Bougainvillia*, *Eubranchus*, *Diopatra*, *Harpacticus*, *Hydractinia*, *Polygordius*, and *Tomopteris*. Still, species-level identifications with 18S should be considered with caution [101,102].

4.3. DNA Metabarcoding Performance in the Detection of Non-Indigenous Species (NIS) in Zooplankton Samples

The complementarity of COI and 18S in species recovery was crucial for the detection of NIS. A high number of NIS (16 species, representing 4.9% of the total detected species) were recovered from multi-spatial and -seasonal sampling in the Lima estuary, where only *Pseudodiaptomus marinus* was detected with both markers. Most of the NIS previously reported, based on historical records, were not detected in the current study, namely *Acartia (Acanthacartia) tonsa* (Dansa, 1849), *Corbicula fluminea* (Müller, 1774), or *Mya arenaria* (Linnaeus, 1758). *Austrominius modestus* was the only NIS previously reported to occur in the Lima estuary (Table S7, Supplementary Material) and recovered through DNA metabarcoding in the current study. Therefore, DNA metabarcoding allowed for the detection of several NIS which appear to be new reports in the Lima estuary, all of which were documented in Portugal, in addition to those already reported from previous metabarcoding-based assessments in a more enclosed site of the estuary: the recreational marina of Viana do Castelo [103–105].

Additionally, several reads recovered from a singular sample were assigned as *Heniochus acuminatus* (Linnaeus, 1758), an Indo-Pacific fish and putative first introduction in the European Atlantic coast. However, morphological records are required for confirmation, but this species has been already documented as a non-indigenous species in both the southwestern Atlantic coast and in the Ukrainian coast of the Black Sea [106–108], and has been more recently reported in Spanish territory (Canary Islands) [109]. On the other hand, the Bryozoa *Watersipora* spp. was detected with COI, and this genus includes the *Watersipora subtorquata* (d'Orbigny, 1852) that has been reported as a NIS in Portugal, as well as the *Watersipora subatra* (Ortmann, 1890) which is a NIS in European waters, but species-level identifications were not possible with any of the genetic markers here employed. Hence, it is possible that either species was herein detected; however, further confirmation is required with, i.e., morphological assessments. The 18S reads further recovered *Balanus* sp. which COI resolved as *B. trigonus*. Furthermore, *Musculus lateralis* (Say, 1822) assigned reads were recovered with 18S, and a further revision posed it as a reliable identification based on current available data. This species is native to the Atlantic coast of North America and, to our knowledge, it has not yet been recognized as a NIS on European coasts, but a recent 18S-based metabarcoding assessment has also reported its presence on the coast of Sweeden; however, further morphological evidence is required to confirm its occurrence on European coasts [110].

Although *Mytilus* sp. COI and 18S reads here recovered were assigned as being *Mytilus edulis* (Linnaeus, 1758), both markers are unable to discriminate *Mytilus* spp. due to the unusual hybridization and biparental mtDNA inheritance [111,112]. Still, although it is not possible to confirm or deny the occurrence of *M. edulis*, such detections may in fact be considered the congener *Mytilus galloprovincialis* (Lamarck, 1819), since the distribution of both species has been well stipulated [113,114].

However, 18S NIS detections must be considered with caution. For further details on the reliability of the herein NIS detections, see Table S8 (Supplementary Material). Indeed, an assignment with an identity percent higher than 97% with COI leads, in general, to correct species identification (with a few exceptions, i.e., *Mytilus* sp.); whereas, in many cases, the assignment of 18S (even at 100% identity) may yield taxa not present in the studied areas due to the fact that related species included in the reference database may share exactly the same sequence for the 18S fragment used [115]. In the current study, three 18S-recovered NIS were considered unreliable assignments, namely the Bivalvia *M. mercenaria* and both Stolidobranchia (Ascidiacea) *M. squamiger* and *S. plicata* (Table S8, Supplementary Material); however, all of them have been reported on the Portuguese coast [116]. So, further effort is crucial in surveying the estuary, which, coupled with morphology-based assessment, would further improve the confirmation of the introduction and establishment of such species.

4.4. Species Richness, Taxonomic Distinctness, and Species Composition Influenced Primarly by Season and Secondarily by Within-Estuary Location

Complementarity between the markers was apparent throughout the study (Figure 2A) and consistent across the seasonal and spatial patterns of the recovered taxa, thus allowing us to attain a more complete picture of the Lima estuary's zooplankton dynamics. Spatial patterns of zooplankton diversity were less evident than seasonal effects (Figures 6 and 7), which are yet to be documented in the Lima estuary. However, similar conclusions were attained when considering abundance/biomass quantification [16,42,82,83].

The zooplankton spatial distribution is primarily associated with the geomorphology of the estuary (i.e., length, width, depth) that influences the hydrodynamics and concurrently physical/chemical and biological factors, which by itself promotes remarkable influence over zooplankton diversity disparity through the spatial gradient. However, seasonal changes are much more dynamic and relevant over zooplankton taxa by further having an effect over spatial parameters. Therefore, zooplankton richness distribution and composition were more strongly modulated by seasonal variation, while also presenting a considerable spatial shifting specific to sampled seasons throughout the estuary. A zooplankton survey downstream of Eiffel's bridge, from 2010 to 2011, has shown abundance peaks during summer and a secondary peak during autumn [16], similar to zooplankton assessments from other sites at the Lima estuary [85–87], which are characteristic of temperate systems [16,117]. Such findings were comparable to those herein: higher zooplankton richness was observed in the summer and next autumn seasons (Figures 3A and 5), as well as a greater zooplankton diversity and taxonomic representation at the river mouth (Figures 3A and 6) [16]. However, the water parameters we measured also displayed considerable salinity and conductivity changes throughout the spatial gradient of the estuary (Table S1, Supplementary Material). This result was not that apparent in the community's analysis; it is probably related to the fact that our water parameter measures were taken on the superficial portion of the water column, whereas the samples for the community analysis were from deeper layers.

Nevertheless, zooplankton richness peaking in summer is usually associated with previous winter floods with higher nutrients and sediment flows into the estuary [16] and accumulates in seasons of lower water currents, such as in the summer, which promote primary production and facilitate nursery areas [118], particularly as observed for meroplankton (e.g., gastropods and polychaetas) (Figure 3) [87]. Summer conditions have also been associated with greater ichthyoplankton representation [42], but our metabarcoding data showed higher richness in the spring. Autumn and spring patterns were demonstrated to be similar to previous abundance/biomass arrangements [85–87]. However, seasonal variation influenced several taxonomic groups differently, which shows that the taxa from different phyla responded differently to seasonal variation, but the majority highlighted the high dissimilarity of LMZ4 communities' composition in comparison to remaining locations. Indeed, this study highlighted particular seasonal divergences between Arthropoda and Annelida and Mollusca and the general zooplankton patterns, indicating similar species-level compositions between summer/spring and autumn/spring for the former two taxonomic groups while the latter two shared higher similarity. For annelids, the taxonomic distinctness appeared to have supported species-level clustering (Figure 6), since the most related seasons demonstrated similar taxonomic distinctness as well. Still, the available data on the Lima estuary's zooplankton is scarce, and partially and/or does not represent our sampling sites.

5. Conclusions

Overall, the present study provided a more resolved analysis of zooplankton occurring seasonally, spanning approximately 6 km of the Lima estuary, using a multi-marker DNA metabarcoding approach. Both the composition and species richness were demonstrated to be differently distributed due to seasonal variation and over the spatial gradient, although with a greater influence from the former. The autumn-occurring zooplankton from the most downstream site (LMZ4) of the estuary displayed the most unique composition, with several relevant species and high Hydrozoa and Pycnogonida richness; however, several dominant phyla responded differently to seasonal and spatial variation. Furthermore, the results demonstrated a high relevance of meroplankton in the zooplankton of the Lima estuary, something that has been also highlighted in previous studies, but with low taxonomic resolution. In addition, several NIS were detected which were not yet reported in previous morphology-based surveys of the Lima estuary, although already reported to occur in Portugal. These findings highlight the need for more studies on zooplankton composition in the estuary and surrounding areas. Such studies are crucial for the improvement of models and ecological quality assessments, supporting conservation and more sustainable ecosystem service management, and mitigating climate change's effects on highly dynamic ecosystems, such as coastal ecosystems.

Supplementary Materials: The following supporting information can be downloaded at: https://www.mdpi.com/article/10.3390/ani13243876/s1, Table S1: Measured physical and chemical parameters at each sampling event, and the specific coordinates for each sampling site; Table S2: Nanodrop results regarding DNA concentrations of all zooplankton samples; Table S3: Number of reads recovered from Illumina MiSeq sequencing and throughout the whole process of filtration, trimming, and taxonomic assignment until eligibility for analysis; Table S4: List of species and respective number of reads recovered with DNA metabarcoding from zooplankton samples using two molecular markers, COI and 18S. Non-indigenous species are marked with *;Table S5: Significant ($p > 0.05$) correlated species to both axis in the nMDS (NMDS1 and NMDS2) for the whole zooplankton recovered data; Table S6: Species list that clustered into 7 groups (G1-7); Table S7: Compiled list of zooplankton (Z) and macrozoobenthos (M) species reported from previous morphology-based studies, as well as markers which detected them in the current study; Table S8. List of the species detected in the current study, with associated marker recovery, and notes on the reliability of the detection; Figure S1: Species-level (y-axis) and sampling events (x-axis) clustering associated with a heatmap color-coded based on the number of replicates where each taxon was detected on each sampling event (0–3, with 0 indicating absence in all replicates and 3 presence in all replicates). Both dendrograms are color-coded based on each taxa phylum for the y-axis and based on the sampled seasons for the x-axis (blue = summer, green = spring, red = autumn). Cluster-specific species are found to the right of the heatmap; Info S1: Genoinseq amplification and sequencing report; Info S2: List of the reference libraries used for taxonomic assignment of COI reads (in order). Refs. [47–49,56,57,119–121] are cited in supplementary files.

Author Contributions: Conceptualization, J.M., P.T.G., F.O.C. and S.D.; Methodology, J.M., S.D., P.T.G., F.O.C. and D.C.-F.; Validation, F.O.C., D.C.-F. and S.D.; Formal analysis, J.M.; Investigation, J.M.; resources, F.O.C., P.T.G. and S.D.; Data curation, J.M.; Writing—original draft, J.M.; Writing—review and editing, J.M., F.O.C., S.D., D.C.-F. and P.T.G.; Visualization, J.M.; Supervision, F.O.C. and S.D.; Project administration, F.O.C. and S.D. All authors have read and agreed to the published version of the manuscript.

Funding: This work was funded by the project "ATLANTIDA: Platform for the monitoring of the North Atlantic Ocean and tools for the sustainable exploitation of the marine resources" (Norte-01-0145-FEDER-000040), funded by the Programa Operacional Regional do Norte (NORTE2020), co-financed by the European Regional Development Fund (ERDF), and by the "Contrato-Programa" UIDB/04050/2020 funded by national funds through the Foundation for Science and Technology, FCT I.P. https://doi.org/10.54499/UIDB/04050/2020. Financial support granted by the FCT to SD (CEECIND/00667/2017) is also acknowledged.

Institutional Review Board Statement: Not applicable.

Informed Consent Statement: Not applicable.

Data Availability Statement: All data generated and analyzed during this study are included in this article and in its Supplementary Materials.

Acknowledgments: The authors want to thank Ana Sofia Lavrador, Pedro Vieira, João Tadeu, André Ferreira, and Cláudia Machado for all the help provided in the lab and in the field. The authors are also grateful to three reviewers for the comments and suggestions that greatly improved the manuscript.

Conflicts of Interest: The authors declare no conflict of interest.

References

1. Hellawell, J.M. *Biological Indicators of Freshwaters Pollution and Environmental Management;* Mellanby, K., Ed.; Elsevier: London, UK; New York, NY, USA, 1986; ISBN 9789400943155.
2. Morse, J.C.; Bae, Y.J.; Munkhjargal, G.; Sangpradub, N.; Tanida, K.; Vshivkova, T.S.; Wang, B.; Yang, L.; Yule, C.M. Freshwater Biomonitoring with Macroinvertebrates in East Asia. *Front. Ecol. Environ.* **2007**, *5*, 33–42. [CrossRef]
3. Ruaro, R.; Gubiani, É.A.; Cunico, A.M.; Moretto, Y.; Piana, P.A. Comparison of Fish and Macroinvertebrates as Bioindicators of Neotropical Streams. *Environ. Monit. Assess.* **2016**, *188*, 1–13. [CrossRef] [PubMed]
4. Caroni, R.; Irvine, K. The Potential of Zooplankton Communities for Ecological Assessment of Lakes: Redundant Concept or Political Oversight? *Biol. Environ.* **2010**, *110*, 35–53. [CrossRef]
5. Landry, M.R.; Hassett, R.P. Estimating the Grazing Impact of Marine Micro-Zooplankton. *Mar. Biol.* **1982**, *67*, 283–288. [CrossRef]
6. Ikeda, T. Nutritional Ecology of Zooplankton. Ph.D. Thesis, Hokkaido University, Hakodate, Japan, 1974.
7. Suthers, I.M.; Rissik, D. *Plankton: A Guide to Their Ecology and Monitoring for Water Quality*, 2nd ed.; CSIRO: Canberra, Australia, 2009; ISBN 9780643090583.
8. Le Quéré, C.; Buitenhuis, E.T.; Moriarty, R.; Alvain, S.; Aumont, O.; Bopp, L.; Chollet, S.; Enright, C.; Franklin, D.J.; Geider, R.J.; et al. Role of Zooplankton Dynamics for Southern Ocean Phytoplankton Biomass and Global Biogeochemical Cycles. *Biogeosciences* **2016**, *13*, 4111–4133. [CrossRef]
9. Steinberg, D. Zooplankton Biogeochemical Cycles. In *Marine Plankton: A Practical Guide to Ecology, Methodology, and Taxonomy;* Castellani, C., Edwards Martin, Eds.; Oxford University Press: Oxford, UK, 2017; Volume 1.
10. Gannon, J.E.; Stemberger, R.S. Zooplankton (Especially Crustaceans and Rotifers) as Indicators of Water Quality. *Trans. Am. Microsc. Soc.* **1978**, *97*, 16–35. [CrossRef]
11. An, X.P.; Du, Z.H.; Zhang, J.H.; Li, Y.P.; Qi, J.W. Structure of the Zooplankton Community in Hulun Lake, China. *Procedia Environ. Sci.* **2012**, *13*, 1099–1109. [CrossRef]
12. Azevêdo, D.J.S.; Barbosa, J.E.L.; Gomes, W.I.A.; Porto, D.E.; Marques, J.C.; Molozzi, J. Diversity Measures in Macroinvertebrate and Zooplankton Communities Related to the Trophic Status of Subtropical Reservoirs: Contradictory or Complementary Responses? *Ecol. Indic.* **2015**, *50*, 135–149. [CrossRef]
13. Kour, S.; Slathia, D.; Sharma, N.; Kour, S.; Verma, R. Zooplankton as Bioindicators of Trophic Status of a Lentic Water Source, Jammu (J&K) with Remarks on First Reports. *Proc. Natl. Acad. Sci. India Sect. B Biol. Sci.* **2022**, *92*, 393–404. [CrossRef]
14. Chiba, S.; Batten, S.; Martin, C.S.; Ivory, S.; Miloslavich, P.; Weatherdon, L.V. Zooplankton Monitoring to Contribute towards Addressing Global Biodiversity Conservation Challenges. *J. Plankton Res.* **2018**, *40*, 509–518. [CrossRef]
15. Almeida, R.; Formigo, N.E.; Sousa-Pinto, I.; Antunes, S.C. Contribution of Zooplankton as a Biological Element in the Assessment of Reservoir Water Quality. *Limnetica* **2020**, *39*, 245–261. [CrossRef]
16. Vieira, L.R.; Guilhermino, L.; Morgado, F. Zooplankton Structure and Dynamics in Two Estuaries from the Atlantic Coast in Relation to Multi-Stressors Exposure. *Estuar. Coast. Shelf Sci.* **2015**, *167*, 347–367. [CrossRef]
17. Berasategui, A.A.; Calliari, D.L.; Amodeo, M.; Spetter, C.V.; Guinder, V.; Biancalana, F. Interannual Changes in Winter-Spring Zooplankton Estuarine Community Forced by Hydroclimatic Variability—With Special Reference to Bioindicator Species Eurytemora Americana. *Mar. Environ. Res.* **2023**, *186*, 105898. [CrossRef]
18. Marques, S.C.; Pardal, M.Â.; Primo, A.L.; Martinho, F.; Falcão, J.; Azeiteiro, U.; Molinero, J.C. Evidence for Changes in Estuarine Zooplankton Fostered by Increased Climate Variance. *Ecosystems* **2018**, *21*, 56–67. [CrossRef]
19. Almeida, L.R.; Costa, I.S.; Eskinazi-Sant'Anna, E.M. Composition and Abundance of Zooplankton Community of an Impacted Estuarine Lagoon in Northeast Brazil. *Braz. J. Biol.* **2012**, *72*, 12–24. [CrossRef] [PubMed]
20. Goswami, S.C. *Zooplankton Methodology, Collection & Identification—A Field Manual*, 1st ed.; Dhargalkar, V.K., Verlecar, X.N., Eds.; National Institute of Oceanography: Goa, India, 2004.
21. Jeppesen, E.; Nõges, P.; Davidson, T.A.; Haberman, J.; Nõges, T.; Blank, K.; Lauridsen, T.L.; Søndergaard, M.; Sayer, C.; Laugaste, R.; et al. Zooplankton as Indicators in Lakes: A Scientific-Based Plea for Including Zooplankton in the Ecological Quality Assessment of Lakes According to the European Water Framework Directive (WFD). *Hydrobiologia* **2011**, *676*, 279–297. [CrossRef]
22. Taberlet, P.; Coissac, E.; Pompanon, F.; Brochmann, C.; Willerslev, E. Towards Next-Generation Biodiversity Assessment Using DNA Metabarcoding. *Mol. Ecol.* **2012**, *21*, 2045–2050. [CrossRef]
23. Creer, S.; Deiner, K.; Frey, S.; Porazinska, D.; Taberlet, P.; Thomas, W.K.; Potter, C.; Bik, H.M. The Ecologist's Field Guide to Sequence-Based Identification of Biodiversity. *Methods Ecol. Evol.* **2016**, *7*, 1008–1018. [CrossRef]
24. Shokralla, S.; Spall, J.L.; Gibson, J.F.; Hajibabaei, M. Next-Generation Sequencing Technologies for Environmental DNA Research. *Mol. Ecol.* **2012**, *21*, 1794–1805. [CrossRef]

25. Bucklin, A.; Lindeque, P.K.; Rodriguez-Ezpeleta, N.; Albaina, A.; Lehtiniemi, M. Metabarcoding of Marine Zooplankton: Prospects, Progress and Pitfalls. *J. Plankton Res.* **2016**, *38*, 393–400. [CrossRef]
26. Brown, E.A.; Chain, F.J.J.; Zhan, A.; MacIsaac, H.J.; Cristescu, M.E. Early Detection of Aquatic Invaders Using Metabarcoding Reveals a High Number of Non-Indigenous Species in Canadian Ports. *Divers. Distrib.* **2016**, *22*, 1045–1059. [CrossRef]
27. Chain, F.J.J.; Brown, E.A.; Macisaac, H.J.; Cristescu, M.E. Metabarcoding Reveals Strong Spatial Structure and Temporal Turnover of Zooplankton Communities among Marine and Freshwater Ports. *Divers. Distrib.* **2016**, *22*, 493–504. [CrossRef]
28. Santoferrara, L.F. Current Practice in Plankton Metabarcoding: Optimization and Error Management. *J. Plankton Res.* **2019**, *41*, 571–582. [CrossRef]
29. Blanco-Bercial, L. Metabarcoding Analyses and Seasonality of the Zooplankton Community at BATS. *Front. Mar. Sci.* **2020**, *7*, 1–16. [CrossRef]
30. Brandão, M.C.; Comtet, T.; Pouline, P.; Cailliau, C.; Blanchet-Aurigny, A.; Sourisseau, M.; Siano, R.; Memery, L.; Viard, F.; Nunes, F. Oceanographic Structure and Seasonal Variation Contribute to High Heterogeneity in Mesozooplankton over Small Spatial Scales. *ICES J. Mar. Sci.* **2021**, *78*, 3288–3302. [CrossRef]
31. Lindeque, P.K.; Parry, H.E.; Harmer, R.A.; Somerfield, P.J.; Atkinson, A. Next Generation Sequencing Reveals the Hidden Diversity of Zooplankton Assemblages. *PLoS ONE* **2013**, *8*, e81327. [CrossRef] [PubMed]
32. Abad, D.; Albaina, A.; Aguirre, M.; Laza-Martínez, A.; Uriarte, I.; Iriarte, A.; Villate, F.; Estonba, A. Is Metabarcoding Suitable for Estuarine Plankton Monitoring? *A Comparative Study with Microscopy. Mar. Biol.* **2016**, *163*, 1–13. [CrossRef]
33. Harvey, J.B.J.; Johnson, S.B.; Fisher, J.L.; Peterson, W.T.; Vrijenhoek, R.C. Comparison of Morphological and next Generation DNA Sequencing Methods for Assessing Zooplankton Assemblages. *J. Exp. Mar. Biol. Ecol.* **2017**, *487*, 113–126. [CrossRef]
34. Schroeder, A.; Stanković, D.; Pallavicini, A.; Gionechetti, F.; Pansera, M.; Camatti, E. DNA Metabarcoding and Morphological Analysis—Assessment of Zooplankton Biodiversity in Transitional Waters. *Mar. Environ. Res.* **2020**, *160*, 104946. [CrossRef]
35. Coguiec, E.; Ershova, E.A.; Daase, M.; Vonnahme, T.R.; Wangensteen, O.S.; Gradinger, R.; Præbel, K.; Berge, J. Seasonal Variability in the Zooplankton Community Structure in a Sub-Arctic Fjord as Revealed by Morphological and Molecular Approaches. *Front. Mar. Sci.* **2021**, *8*, 705042. [CrossRef]
36. Bucklin, A.; Batta-Lona, P.G.; Questel, J.M.; Wiebe, P.H.; Richardson, D.E.; Copley, N.J.; O'Brien, T.D. COI Metabarcoding of Zooplankton Species Diversity for Time-Series Monitoring of the NW Atlantic Continental Shelf. *Front. Mar. Sci.* **2022**, *9*, 867893. [CrossRef]
37. Bucklin, A.; Yeh, H.D.; Questel, J.M.; Richardson, D.E.; Reese, B.; Copley, N.J.; Wiebe, P.H. Time-Series Metabarcoding Analysis of Zooplankton Diversity of the NW Atlantic Continental Shelf. *ICES J. Mar. Sci.* **2019**, *76*, 1162–1176. [CrossRef]
38. Ershova, E.A.; Wangensteen, O.S.; Descoteaux, R.; Barth-Jensen, C.; Præbel, K. Metabarcoding as a Quantitative Tool for Estimating Biodiversity and Relative Biomass of Marine Zooplankton. *ICES J. Mar. Sci.* **2021**, *78*, 3342–3355. [CrossRef]
39. Zheng, L.; He, J.; Lin, Y.; Cao, W.; Zhang, W. 16S RRNA Is a Better Choice than COI for DNA Barcoding Hydrozoans in the Coastal Waters of China. *Acta Oecol.* **2014**, *33*, 55–76. [CrossRef]
40. Zhang, G.K.; Chain, F.J.J.; Abbott, C.L.; Cristescu, M.E. Metabarcoding Using Multiplexed Markers Increases Species Detection in Complex Zooplankton Communities. *Evol. Appl.* **2018**, *11*, 1901–1914. [CrossRef] [PubMed]
41. Carroll, E.L.; Gallego, R.; Sewell, M.A.; Zeldis, J.; Ranjard, L.; Ross, H.A.; Tooman, L.K.; O'Rorke, R.; Newcomb, R.D.; Constantine, R. Multi-Locus DNA Metabarcoding of Zooplankton Communities and Scat Reveal Trophic Interactions of a Generalist Predator. *Sci. Rep.* **2019**, *9*, 281. [CrossRef] [PubMed]
42. Ramos, S.; Cowen, R.K.; Ré, P.; Bordalo, A.A. Temporal and Spatial Distributions of Larval Fish Assemblages in the Lima Estuary (Portugal). *Estuar. Coast. Shelf Sci.* **2006**, *66*, 303–314. [CrossRef]
43. Ramos, S.; Cowen, R.K.; Paris, C.; Ré, P.; Bordalo, A.A. Environmental Forcing and Larval Fish Assemblage Dynamics in the Lima River Estuary (Northwest Portugal). *J. Plankton Res.* **2006**, *28*, 275–286. [CrossRef]
44. Sousa, R.; Dias, S.; Antunes, J.C. Spatial Subtidal Macrobenthic Distribution in Relation to Abiotic Conditions in the Lima Estuary, NW of Portugal. *Hydrobiologia* **2006**, *559*, 135–148. [CrossRef]
45. Costa-Dias, S.; Sousa, R.; Antunes, C. Ecological Quality Assessment of the Lower Lima Estuary. *Mar. Pollut. Bull.* **2010**, *61*, 234–239. [CrossRef]
46. Azevedo, I.; Ramos, S.; Mucha, A.P.; Bordalo, A.A. Applicability of Ecological Assessment Tools for Management Decision-Making: A Case Study from the Lima Estuary (NW Portugal). *Ocean Coast. Manag.* **2013**, *72*, 54–63. [CrossRef]
47. Leray, M.; Yang, J.Y.; Meyer, C.P.; Mills, S.C.; Agudelo, N.; Ranwez, V.; Boehm, J.T.; Machida, R.J. A New Versatile Primer Set Targeting a Short Fragment of the Mitochondrial COI Region for Metabarcoding Metazoan Diversity: Application for Characterizing Coral Reef Fish Gut Contents. *Front. Zool.* **2013**, *10*, 34. [CrossRef] [PubMed]
48. Lobo, J.; Costa, P.M.; Teixeira, M.A.L.; Ferreira, M.S.G.; Costa, M.H.; Costa, F.O. Enhanced Primers for Amplification of DNA Barcodes from a Broad Range of Marine Metazoans. *BMC Ecol.* **2013**, *13*, 34. [CrossRef] [PubMed]
49. Stoeck, T.; Bass, D.; Nebel, M.; Christen, R.; Jones, M.D.M.; Breiner, H.W.; Richards, T.A. Multiple Marker Parallel Tag Environmental DNA Sequencing Reveals a Highly Complex Eukaryotic Community in Marine Anoxic Water. *Mol. Ecol.* **2010**, *19*, 21–31. [CrossRef] [PubMed]
50. Leite, B.R.; Vieira, P.E.; Troncoso, J.S.; Costa, F.O. Comparing Species Detection Success between Molecular Markers in DNA Metabarcoding of Coastal Macroinvertebrates. *Metabarcoding Metagenom.* **2021**, *5*, 249–260. [CrossRef]

51. Hollatz, C.; Leite, B.R.; Lobo, J.; Froufe, H.; Egas, C.; Costa, F.O. Priming of a DNA Metabarcoding Approach for Species Identification and Inventory in Marine Macrobenthic Communities. *Genome* **2017**, *60*, 260–271. [CrossRef] [PubMed]
52. Clarke, L.J.; Beard, J.M.; Swadling, K.M.; Deagle, B.E. Effect of Marker Choice and Thermal Cycling Protocol on Zooplankton DNA Metabarcoding Studies. *Ecol. Evol.* **2017**, *7*, 873–883. [CrossRef]
53. Pochon, X.; Bott, N.J.; Smith, K.F.; Wood, S.A. Evaluating Detection Limits of Next-Generation Sequencing for the Surveillance and Monitoring of International Marine Pests. *PLoS ONE* **2013**, *8*, e73935. [CrossRef]
54. Zhan, A.; Hulák, M.; Sylvester, F.; Huang, X.; Adebayo, A.A.; Abbott, C.L.; Adamowicz, S.J.; Heath, D.D.; Cristescu, M.E.; Macisaac, H.J. High Sensitivity of 454 Pyrosequencing for Detection of Rare Species in Aquatic Communities. *Methods Ecol. Evol.* **2013**, *4*, 558–565. [CrossRef]
55. van der Loos, L.M.; Nijland, R. Biases in Bulk: DNA Metabarcoding of Marine Communities and the Methodology Involved. *Mol. Ecol.* **2021**, *30*, 3270–3288. [CrossRef]
56. Comeau, A.M.; Douglas, G.M.; Langille, M.G.I. Microbiome Helper: A Custom and Streamlined Workflow for Microbiome Research. *mSystems* **2017**, *2*, 1–11. [CrossRef] [PubMed]
57. Schmieder, R.; Edwards, R. Quality Control and Preprocessing of Metagenomic Datasets. *Bioinformatics* **2011**, *27*, 863–864. [CrossRef] [PubMed]
58. Schloss, P.D.; Westcott, S.L.; Ryabin, T.; Hall, J.R.; Hartmann, M.; Hollister, E.B.; Lesniewski, R.A.; Oakley, B.B.; Parks, D.H.; Robinson, C.J.; et al. Introducing Mothur: Open-Source, Platform-Independent, Community-Supported Software for Describing and Comparing Microbial Communities. *Appl. Environ. Microbiol.* **2009**, *75*, 7537–7541. [CrossRef] [PubMed]
59. Kozich, J.J.; Westcott, S.L.; Baxter, N.T.; Highlander, S.K.; Schloss, P.D. Development of a Dual-Index Sequencing Strategy and Curation Pipeline for Analyzing Amplicon Sequence Data on the Miseq Illumina Sequencing Platform. *Appl. Environ. Microbiol.* **2013**, *79*, 5112–5120. [CrossRef] [PubMed]
60. Ratnasingham, S. mBRAVE: The Multiplex Barcode Research And Visualization Environment. *Biodivers. Inf. Sci. Stand.* **2019**, 3. [CrossRef]
61. Quast, C.; Pruesse, E.; Yilmaz, P.; Gerken, J.; Schweer, T.; Yarza, P.; Peplies, J.; Glöckner, F.O. The SILVA Ribosomal RNA Gene Database Project: Improved Data Processing and Web-Based Tools. *Nucleic Acids Res.* **2013**, *41*, 590–596. [CrossRef] [PubMed]
62. Ratnasingham, S.; Hebert, P.D.N. BOLD: The Barcode of Life Data System: Barcoding. *Mol. Ecol. Notes* **2007**, *7*, 355–364. [CrossRef]
63. Pruesse, E.; Peplies, J.; Glöckner, F.O. SINA: Accurate High-Throughput Multiple Sequence Alignment of Ribosomal RNA Genes. *Bioinformatics* **2012**, *28*, 1823–1829. [CrossRef]
64. Rognes, T.; Flouri, T.; Nichols, B.; Quince, C.; Mahé, F. VSEARCH: A Versatile Open Source Tool for Metagenomics. *PeerJ* **2016**, *2016*, e2584. [CrossRef]
65. Camacho, C.; Coulouris, G.; Avagyan, V.; Ma, N.; Papadopoulos, J.; Bealer, K.; Madden, T.L. BLAST+: Architecture and Applications. *BMC Bioinform.* **2009**, *10*, 421. [CrossRef]
66. Duarte, S.; Vieira, P.E.; Leite, B.R.; Teixeira, M.A.L.; Neto, J.M.; Costa, F.O. Macrozoobenthos Monitoring in Portuguese Transitional Waters in the Scope of the Water Framework Directive Using Morphology and DNA Metabarcoding. *Estuar. Coast. Shelf Sci.* **2023**, *281*, 108207. [CrossRef]
67. Fais, M.; Duarte, S.; Vieira, P.E.; Sousa, R.; Hajibabaei, M.; Canchaya, C.A.; Costa, F.O. Small-Scale Spatial Variation of Meiofaunal Communities in Lima Estuary (NW Portugal) Assessed through Metabarcoding. *Estuar. Coast. Shelf Sci.* **2020**, *238*, 106683. [CrossRef]
68. Oksanen, J.; Simpson, G.L.; Blanchet, F.G.; Kindt, R.; Legendre, P.; Minchin, P.R.; O'Hara, R.B.; Solymos, P.; Stevens, M.H.H.; Szoecs, E.; et al. Vegan: Community Ecology Package - R Package Version 2.6-4. *CRAN's Repos.* **2022**.
69. Tennekes, M.; Ellis, P. Treemap Visualization. *CRAN's Repos.* **2022**.
70. Clarke, K.R.; Warwick, R.M. A Taxonomic Distinctness Index and Its Statistical Properties. *J. Appl. Ecol.* **1998**, *35*, 523–531. [CrossRef]
71. Kolde, R. Pretty Heatmaps. *CRAN's Repos.* **2022**.
72. Heberle, H.; Meirelles, G.V.; Silva, F.R.; Telles, G.P.; Minghim, R. InteractiVenn: A Web-Based Tool for the Analysis of Sets through Venn Diagrams. *BMC Bioinform.* **2015**, *16*, 7. [CrossRef]
73. Radulovici, A.E.; Vieira, P.E.; Duarte, S.; Teixeira, M.A.L.; Borges, L.M.S.; Deagle, B.E.; Majaneva, S.; Redmond, N.; Schultz, J.A.; Costa, F.O. Revision and Annotation of DNA Barcode Records for Marine Invertebrates: Report of the 8th IBOL Conference Hackathon. *Metabarcoding Metagenom.* **2021**, *5*, 207–217. [CrossRef]
74. Lobo, J.; Teixeira, M.A.L.; Borges, L.M.S.; Ferreira, M.S.G.; Hollatz, C.; Gomes, P.T.; Sousa, R.; Ravara, A.; Costa, M.H.; Costa, F.O. Starting a DNA Barcode Reference Library for Shallow Water Polychaetes from the Southern European Atlantic Coast. *Mol. Ecol. Resour.* **2016**, *16*, 298–313. [CrossRef]
75. Lobo, J.; Ferreira, M.S.; Antunes, I.C.; Teixeira, M.A.L.; Borges, L.M.S.; Sousa, R.; Gomes, P.A.; Helena Costa, M.; Cunha, M.R.; Costa, F.O. Contrasting Morphological and DNA Barcode-Suggested Species Boundaries among Shallow-Water Amphipod Fauna from the Southern European Atlantic Coast. *Genome* **2017**, *60*, 147–157. [CrossRef]
76. Leite, B.R.; Vieira, P.E.; Teixeira, M.A.L.; Lobo-Arteaga, J.; Hollatz, C.; Borges, L.M.S.; Duarte, S.; Troncoso, J.S.; Costa, F.O. Gap-Analysis and Annotated Reference Library for Supporting Macroinvertebrate Metabarcoding in Atlantic Iberia. *Reg. Stud. Mar. Sci.* **2020**, *36*, 101307. [CrossRef]

77. Borges, L.M.S.; Hollatz, C.; Lobo, J.; Cunha, A.M.; Vilela, A.P.; Calado, G.; Coelho, R.; Costa, A.C.; Ferreira, M.S.G.; Costa, M.H.; et al. With a Little Help from DNA Barcoding: Investigating the Diversity of Gastropoda from the Portuguese Coast. *Sci. Rep.* **2016**, *6*, 20226. [CrossRef] [PubMed]
78. Fontes, J.T.; Vieira, P.E.; Ekrem, T.; Soares, P.; Costa, F.O. BAGS: An Automated Barcode, Audit & Grade System for DNA Barcode Reference Libraries. *Mol. Ecol.* **2021**, *21*, 573–583. [CrossRef]
79. Lavrador, A.S.; Fontes, J.T.; Vieira, P.E.; Costa, F.O.; Duarte, S. Compilation, Revision, and Annotation of DNA Barcodes of Marine Invertebrate Non-Indigenous Species (NIS) Occurring in European Coastal Regions. *Diversity* **2023**, *15*, 174. [CrossRef]
80. Oliveira, L.M.; Knebelsberger, T.; Landi, M.; Soares, P.; Raupach, M.J.; Costa, F.O. Assembling and Auditing a Comprehensive DNA Barcode Reference Library for European Marine Fishes. *J. Fish Biol.* **2016**, *89*, 2741–2754. [CrossRef] [PubMed]
81. Moutinho, J. DNA Metabarcoding Monitoring of Zooplankton for the Detection of Non-Indigenous Species (NIS): A Seasonal Study in a Recreational Marina of the Northwest of Portugal. Master's Thesis, University of Minho, Braga, Portugal, 2022.
82. Ramos, S.; Cabral, H.; Elliott, M. Do Fish Larvae Have Advantages over Adults and Other Components for Assessing Estuarine Ecological Quality? *Ecol. Indic.* **2015**, *55*, 74–85. [CrossRef]
83. Ramos, S.; Ré, P.; Bordalo, A.A. Recruitment of Flatfish Species to an Estuarine Nursery Habitat (Lima Estuary, NW Iberian Peninsula). *J. Sea Res.* **2010**, *64*, 473–486. [CrossRef]
84. Ramos, S. *Ichthyoplankton of the Lima Estuary (NW Portugal): Ecology of the Early Life Stages of Pleuronectiformes*; University of Porto: Porto, Portugal, 2007.
85. Guimarães, C.; Galhano, H. Ecological Study of the Estuary of River Lima (Portugal): I—The North Bank Saltmarshes. In *Publicações do Instituito de Zoologia "Dr. Augusto Nobre"*; Faculdade de Ciências do Porto: Porto, Portugal, 1987; pp. 1–54.
86. Guimarães, C.; Galhano, H. Ecological Study of the Estuary of River Lima (Portugal): II—A Mud-Sandybeach. In *Publicações do Instituito de Zoologia "Dr. Augusto Nobre"*; Faculdade de Ciências do Porto: Porto, Portugal, 1988; pp. 1–73.
87. Guimarães, C.; Galhano, H. Ecological Study of the Estuary of River Lima (Portugal): III—Channels of Darque. In *Publicações do Instituito de Zoologia "Dr. Augusto Nobre"*; Faculdade de Ciências do Porto: Porto, Portugal, 1989; pp. 1–52.
88. Stefanni, S.; Stanković, D.; Borme, D.; de Olazabal, A.; Juretić, T.; Pallavicini, A.; Tirelli, V. Multi-Marker Metabarcoding Approach to Study Mesozooplankton at Basin Scale. *Sci. Rep.* **2018**, *8*, 12085. [CrossRef]
89. Belmonte, G.; Rubino, F. Resting Cysts from Coastal Marine Plankton. In *Oceanography and Marine Biology an Annual Review*; Hawkins, S.J., Allcock, A.L., Bates, A.E., Firth, L.B., Smith, I.P., Swearer, S.E., Todd, P.A., Eds.; CRC Press: Boca Raton, FL, USA; Taylor & Francis Group: Milton Park, UK, 2019; Volume 57, pp. 1–88.
90. Belmonte, G.; Vaglio, I.; Rubino, F.; Alabiso, G. Zooplankton Composition along the Confinement Gradient of the Taranto Sea System (Ionian Sea, South-Eastern Italy). *J. Mar. Syst.* **2013**, *128*, 222–238. [CrossRef]
91. Rubino, F.; Belmonte, G. Habitat Shift for Plankton: The Living Side of Benthic-Pelagic Coupling in the Mar Piccolo of Taranto (Southern Italy, Ionian Sea). *Water* **2021**, *13*, 3619. [CrossRef]
92. Intxausti, L.; Villate, F.; Uriarte, I.; Iriarte, A.; Ameztoy, I. Size-Related Response of Zooplankton to Hydroclimatic Variability and Water-Quality in an Organically Polluted Estuary of the Basque Coast (Bay of Biscay). *J. Mar. Syst.* **2012**, *94*, 87–96. [CrossRef]
93. Sousa, R. *Estrutura Das Comunidades de Macroinvertebrados Bentónicos Presentes No Estuário Do Rio Lima*; University of Porto: Porto, Portugal, 2003.
94. Largier, J.; Delgadillo, F.; Grierson, P. Seasonally Hypersaline Estuaries in Mediterranean-Climate Regions. *Estuar. Coast. Shelf Sci.* **1997**, *45*, 789–797. [CrossRef]
95. Valente, A.C.N.; Alexandrino, P.J.B. Ecological Study of the Estuary of River Lima. IV. The Ichthyofauna in the Darque Channels (River Lima Estuary) with Special Reference to the Biology of the Sand-Melt, Atherina presbyter Cuvier, 1829 (Pisces: Atherinidae). *Publicações do Instituto de Zoologia "Dr. Augusto Nobre"* **1988**, *202*, 1–17.
96. Hebert, P.D.N.; Ratnasingham, S.; DeWaard, J.R. Barcoding Animal Life: Cytochrome c Oxidase Subunit 1 Divergences among Closely Related Species. *Proc. R. Soc. B Biol. Sci.* **2003**, *270*, S96–S99. [CrossRef] [PubMed]
97. Capra, E.; Giannico, R.; Montagna, M.; Turri, F.; Cremonesi, P.; Strozzi, F.; Leone, P.; Gandini, G.; Pizzi, F. A New Primer Set for DNA Metabarcoding of Soil Metazoa. *Eur. J. Soil Biol.* **2016**, *77*, 53–59. [CrossRef]
98. Mueller, R.L. Evolutionary Rates, Divergence Dates, and the Performance of Mitochondrial Genes in Bayesian Phylogenetic Analysis. *Syst. Biol.* **2006**, *55*, 289–300. [CrossRef]
99. Gouy, M.; Li, W.H. Molecular Phylogeny of the Kingdoms Animalia, Plantae, and Fungi. *Mol. Biol. Evol.* **1989**, *6*, 109–122. [CrossRef]
100. Amaral-Zettler, L.A.; McCliment, E.A.; Ducklow, H.W.; Huse, S.M. A Method for Studying Protistan Diversity Using Massively Parallel Sequencing of V9 Hypervariable Regions of Small-Subunit Ribosomal RNA Genes. *PLoS ONE* **2009**, *4*, e6372. [CrossRef]
101. Tang, C.Q.; Leasi, F.; Obertegger, U.; Kieneke, A.; Barraclough, T.G.; Fontaneto, D. The Widely Used Small Subunit 18S RDNA Molecule Greatly Underestimates True Diversity in Biodiversity Surveys of the Meiofauna. *Proc. Natl. Acad. Sci. USA* **2012**, *109*, 16208–16212. [CrossRef]
102. Questel, J.M.; Hopcroft, R.R.; DeHart, H.M.; Smoot, C.A.; Kosobokova, K.N.; Bucklin, A. Metabarcoding of Zooplankton Diversity within the Chukchi Borderland, Arctic Ocean: Improved Resolution from Multi-Gene Markers and Region-Specific DNA Databases. *Mar. Biodivers.* **2021**, *51*, 1–19. [CrossRef]

103. Moutinho, J.; Lavrador, A.S.; Vieira, P.E.; Costa, F.O.; Duarte, S. Assessing the Seasonal Dynamics of Zooplankton in a Recreational Marina of the Northwest of Portugal through Multi-Marker DNA Metabarcoding. In *Proceedings of the ARPHA Conference Abstracts*; Pensoft Publishers: Sofia, Bulgaria, 2022; Volume 5.
104. Lavrador, A.S.; Amaral, F.G.; Moutinho, J.; Vieira, P.E.; Costa, F.O.; Duarte, S. Detection and Monitoring of Non-Indigenous Invertebrate Species in Recreational Marinas through DNA Metabarcoding of Zooplankton Communities in the North of Portugal. In Proceedings of the MetaZooGene Symposium: New Insights into Biodiversity, Biogeography, Ecology, and Evolution of Marine Zooplankton Based on Molecular Approaches, Dublin, Ireland, 23 September 2022. Available online: https://metazoogene.org/symposium2022 (accessed on 12 December 2023).
105. Lavrador, A.; Amaral, F.; Vieira, P.E.; Costa, F.; Duarte, S. Surveillance of Non-Indigenous Invertebrate Species through DNA Metabarcoding in Recreational Marinas in the North and Center of Portugal. In *Proceedings of the ARPHA Conference Abstracts*; Pensoft Publishers: Sofia, Bulgaria, 2021; Volume 4.
106. Katsanevakis, S.; Bogucarskis, K.; Gatto, F.; Vandekerkhove, J.; Deriu, I.; Cardoso, A.C. Building the European Alien Species Information Network (EASIN): A Novel Approach for the Exploration of Distributed Alien Species Data. *Bioinvasions Rec.* **2012**, *1*, 235–245. [CrossRef]
107. Luiz, O.J.; Comin, E.J.; Madin, J.S. Far Away from Home: The Occurrence of the Indo-Pacific Bannerfish Heniochus Acuminatus (Pisces: Chaetodontidae) in the Atlantic. *Bull. Mar. Sci.* **2014**, *90*, 741–744. [CrossRef]
108. Adelir-Alves, J.; Soeth, M.; Braga, R.R.; Spach, H.L. Non-Native Reef Fishes in the Southwest Atlantic Ocean: A Recent Record of Heniochus Acuminatus (Linnaeus, 1758) (Perciformes, Chaetodontidae) and Biological Aspects of Chromis Limbata (Valenciennes, 1833) (Perciformes, Pomacentridae). *Check List* **2018**, *14*, 379–385. [CrossRef]
109. Png-Gonzalez, L.; Comas-González, R.; Calvo-Manazza, M.; Follana-Berná, G.; Ballesteros, E.; Díaz-Tapia, P.; Falcón, J.M.; García Raso, J.E.; Gofas, S.; González-Porto, M.; et al. Updating the National Baseline of Non-Indigenous Species in Spanish Marine Waters. *Diversity* **2023**, *15*, 630. [CrossRef]
110. Obst, M. *18S Metabarcoding Genetic Observations of Marine Species in the Port of Wallhamn, Sweden (2022)*; University of Gothenburg: Göteborg, Sweden, 2023.
111. Hoeh, W.R.; Blakley, K.H.; Brown, W.M. Heteroplasmy Suggests Limited Biparental Inheritance of Mytilus Mitochondrial DNA. *Science (1979)* **1991**, *251*, 1488–1490. [CrossRef] [PubMed]
112. Śmietanka, B.; Burzyński, A.; Hummel, H.; Wenne, R. Glacial History of the European Marine Mussels Mytilus, Inferred from Distribution of Mitochondrial DNA Lineages. *Heredity* **2014**, *113*, 250–258. [CrossRef] [PubMed]
113. Boukadida, K.; Mlouka, R.; Clerandeau, C.; Banni, M.; Cachot, J. Natural Distribution of Pure and Hybrid Mytilus Sp. along the South Mediterranean and North-East Atlantic Coasts and Sensitivity of D-Larvae Stages to Temperature Increases and Metal Pollution. *Sci. Total Environ.* **2021**, *756*, 143675. [CrossRef]
114. Hilbish, T.J.; Mullinax, A.; Dolven, S.I.; Meyer, A.; Koehn, R.K.; Rawson, P.D. Origin of the Antitropical Distribution Pattern in Marine Mussels (Mytilus Spp.): Routes and Timing of Transequatorial Migration. *Mar. Biol.* **2000**, *136*, 69–77. [CrossRef]
115. Wangensteen, O.S.; Palacín, C.; Guardiola, M.; Turon, X. DNA Metabarcoding of Littoral Hardbottom Communities: High Diversity and Database Gaps Revealed by Two Molecular Markers. *PeerJ* **2018**, *2018*, e4705. [CrossRef]
116. Chainho, P.; Fernandes, A.; Amorim, A.; Ávila, S.P.; Canning-Clode, J.; Castro, J.J.; Costa, A.C.; Costa, J.L.; Cruz, T.; Gollasch, S.; et al. Non-Indigenous Species in Portuguese Coastal Areas, Coastal Lagoons, Estuaries and Islands. *Estuar. Coast. Shelf Sci.* **2015**, *167*, 199–211. [CrossRef]
117. Marques, S.C.; Pardal, M.A.; Pereira, M.J.; Gonçalves, F.; Marques, J.C.; Azeiteiro, U.M. Zooplankton Distribution and Dynamics in a Temperate Shallow Estuary. *Hydrobiologia* **2007**, *587*, 213–223. [CrossRef]
118. Morais, P.; Chícharo, M.A.; Chícharo, L. Changes in a Temperate Estuary during the Filling of the Biggest European Dam. *Sci. Total Environ.* **2009**, *407*, 2245–2259. [CrossRef] [PubMed]
119. Illumina. *16S Metagenomic Sequencing Library Preparation Preparing 16S Ribosomal RNA Gene Amplicons for the Illumina MiSeq System*; Illumina Technical Document, (Part. No. 15044223 Rev. B.); Illumina: Minato, Tokyo, 2013.
120. Lejzerowicz, F.; Esling, P.; Pillet, L.; Wilding, T.A.; Black, K.D.; Pawlowski, J. High-throughput sequencing and morphology perform equally well for benthic monitoring of marine ecosystems. *Sci. Rep.* **2015**, *5*, 13932. [CrossRef] [PubMed]
121. Schubert, M.; Lindgreen, S.; Orlando, L. AdapterRemoval v2: Rapid adapter trimming, identification, and read merging Findings Background. *BMC Res. Notes* **2016**, *9*, 88. [CrossRef]

animals MDPI

Article

Small-Mammal Genomics Highlights Viaducts as Potential Dispersal Conduits for Fragmented Populations

Tabitha C. Y. Hui [1,2,*], Qian Tang [2], Elize Y. X. Ng [2,3], Ju Lian Chong [4], Eleanor M. Slade [1] and Frank E. Rheindt [2]

1 Asian School of the Environment, Nanyang Technological University, 50 Nanyang Avenue, Singapore 639798, Singapore
2 Department of Biological Sciences, National University of Singapore, 16 Science Drive 4, Singapore 117558, Singapore; dbstq@nus.edu.sg (Q.T.); elizeng@u.nus.edu (E.Y.X.N.); dbsrfe@nus.edu.sg (F.E.R.)
3 Discipline of Biological Sciences, School of Natural Sciences, University of Tasmania, Hobart, TAS 7005, Australia
4 Faculty of Science and Marine Environment, Universiti Malaysia Terengganu, Kuala Terengganu 21030, Terengganu, Malaysia; julian@umt.edu.my
* Correspondence: tabitha.hui@gmail.com

Simple Summary: Wildlife crossings are often constructed to enhance genetic connectivity among populations divided by roads (including highways). However, few studies have demonstrated the efficacy of viaducts in counteracting the barrier effects imposed by roads. We measured genetic diversity and divergence in four small mammal species commonly found in rainforests in Malaysia—*Tupaia glis*, *Maxomys rajah*, *M. whiteheadi*, and *Niviventer cremoriventer*—across three treatment types: (1) viaduct sites, at which sampling locations were separated by a highway but connected by a vegetated viaduct; (2) non-viaduct sites, at which sampling locations were separated by a highway and not connected by a viaduct; and (3) control sites, at which there was no road or highway fragmenting the forest. We found that viaducts facilitated movement in small ground-dwelling species such as *M. whiteheadi* and also when existing highways were relatively wide. However, despite the potential for viaducts to facilitate movement and therefore increase genetic connectivity in *M. whiteheadi*, the genetic distance in populations at viaduct sites was still greater than at control and/or non-viaduct sites for the other three species. Our findings highlight the importance of maintaining intact forests rather than relying solely on the construction of viaducts to connect fragmented populations.

Abstract: Wildlife crossings are implemented in many countries to facilitate the dispersal of animals among habitats fragmented by roads. However, the efficacy of different types of habitat corridors remains poorly understood. We used a comprehensive sampling regime in two lowland dipterocarp forest areas in peninsular Malaysia to sample pairs of small mammal individuals in three treatment types: (1) viaduct sites, at which sampling locations were separated by a highway but connected by a vegetated viaduct; (2) non-viaduct sites, at which sampling locations were separated by a highway and not connected by a viaduct; and (3) control sites, at which there was no highway fragmenting the forest. For four small mammal species, the common tree shrew *Tupaia glis*, Rajah's spiny rat *Maxomys rajah*, Whitehead's spiny rat *Maxomys whiteheadi* and dark-tailed tree rat *Niviventer cremoriventer*, we used genome-wide markers to assess genetic diversity, gene flow and genetic structure. The differences in genetic distance across sampling settings among the four species indicate that they respond differently to the presence of highways and viaducts. Viaducts connecting forests separated by highways appear to maintain higher population connectivity than forest fragments without viaducts, at least in *M. whiteheadi*, but apparently not in the other species.

Keywords: ddRADseq; genetic connectivity; fragmentation; Southeast Asia; local population extinction

1. Introduction

Habitat fragmentation, degradation, and loss pose the most significant threats to the structure and persistence of animal populations and communities [1,2]. Fragmentation is most rapid in developing countries where the expansion of road networks is increasing due to competing land uses such as farming, manufacturing and housing [3]. Barriers that bisect continuous habitat, particularly roads (including highways), initiate the process of habitat fragmentation and can restrict or eliminate animal movement through a landscape [4], with concomitant consequences for connectivity and gene flow [5]. The reduced connectivity may lead to a decrease in viability and persistence of isolated populations [6]. However, the extent to which habitat fragmentation has a negative effect on the genetic structure and persistence of animal populations remains debated. While some studies have demonstrated a negative impact of habitat fragmentation on species' population genetic structure [5,7], others have failed to detect these effects [8].

Wildlife corridors are widely understood to connect habitat fragments and mediate the effects of fragmentation [9]. They include naturally occurring linear habitats such as riparian reserves and hedgerows, purpose-built structures such as wildlife overpasses and underpasses, and incidental structures such as drainage culverts. Corridors have been widely advocated as essential components of reserve design because they can connect isolated areas of suitable habitat and thus minimise the harmful effects of habitat fragmentation on animal movement [10–13]. Corridors are predicted to benefit populations in patchy habitats by promoting movement, which increases population densities, gene flow, and recolonisation of extinct patch populations [14]. However, the efficacy of such passages remains largely untested [15–17], and there has been much debate about their effectiveness in connecting isolated populations [18,19]. Corridors have been shown to increase connectivity, maintain biodiversity [20], increase population sizes [21–23], facilitate movement between fragmented patches [10,18,20,24–30] and promote gene flow [28,31,32]. Other studies, however, have found no significant effects of corridors or even negative impacts [33–35].

To demonstrate the effectiveness of wildlife corridors in increasing connectivity between populations, studies of genetic relatedness are recommended. Molecular techniques such as next-generation sequencing, which has recently been used to reveal fine-scale population structure [36,37], offer new promise in investigating the influence of fragmentation and barriers on population connectivity [38,39]. Genetic methods allow us to measure average migration rates over time, which reveals the effects of fragmentation over several generations and is not as sensitive to current population sizes as mark-recapture studies are (e.g., when populations are extremely low, mark-recapture studies may be impossible) [40]. In addition, molecular techniques measure effective dispersal, the amount of gene flow between populations [41]. As genetic techniques use a single temporal sample per population to estimate migration rather than multiple samples, these techniques require less field effort than mark-recapture [28].

The consequences of habitat fragmentation on dispersal and genetic diversity are still largely unknown for non-volant small mammals, especially in some of the world's equatorial rainforest areas, such as peninsular Malaysia in Southeast Asia. The relative scarcity of data on the effects of fragmentation on small mammals in Southeast Asia is alarming, as these animals provide important ecosystem functions and services for their natural habitats [42]. They are important seed dispersers, pollinators, invertebrate and seed predators, as well as prey for larger predators. Small mammal communities provide a good model for studying such impacts because species in these communities generally use a wide variety of resources, have short generation times that allow for quick detection of environmental change, may be permanent residents of a site, and usually respond to disturbances in a perceptible and measurable way [43]. Smaller mammals are thought to be particularly susceptible to fragmentation due to their limited ability to travel over long distances through exposed habitats [44]. They are thus likely to suffer severe impacts of fragmentation and would benefit from any increases in connectivity brought about by the construction of wildlife corridors.

In this study, we assessed the constraints in migration and gene flow due to habitat fragmentation caused by the construction of highways, and the effectiveness of wildlife underpasses, known in Malaysia as eco-viaducts, in facilitating movement and genetic connectivity in small mammals by comparing genetic distances (a measure of genetic differences, computed by using allele frequency data from many different loci) between individuals at viaduct sites, non-viaduct sites and control sites. We assumed that genetic relationships at all sites would have been at similar levels of divergence prior to fragmentation. We hypothesized that: (1) for forest species, the genetic distance between individuals should be lowest for populations in intact forest and highest for populations fragmented by highways; (2) if eco-viaducts are effective in maintaining population connectivity, the genetic distance between individuals connected by an eco-viaduct will more closely resemble the distance between individuals in intact forest; (3) alternatively, if eco-viaducts are not effective, then pairwise genetic distances should be similar to those in pairs separated by highways. By examining the population genetic structure of small mammal species across different spatial settings in this study, we aim to determine the effectiveness of eco-viaducts across highways in maintaining genetic linkage in a fragmented landscape.

2. Materials and Methods

2.1. Study Sites

We conducted this study in Kenyir, Terengganu (Figure 1a, elevation 100–300 m) and Sungai Yu, Pahang (Figure 1b, elevation 130–210 m), Peninsular Malaysia. Terrain is hilly and consists mostly of lowland dipterocarp forest. Rainfall averages 3000 mm per year with a pronounced wet season from November to March. Flooding is common during this period. The study areas are gazetted as forest reserves and can be logged under permit. Kenyir and Sungai Yu adjoin the Taman Negara National Park, Malaysia's first national park, to the north and west, respectively (Figure 1c). Kenyir and Sungai Yu have rich biodiversity, but are prone to illegal logging, conversion to plantations, and poaching as they do not have the same protection status as a national park. Kenyir forest is bisected by federal route 185 (henceforth highway 185) (Figure 1a) while Sungai Yu forest is bisected by federal route 34 (henceforth highway 34) (Figure 1b). Highway 185 is a two-lane single carriageway (one lane in each direction) with a width of about 8 m and shoulder width of 2 m on either side. Highway 34 is a four-lane dual carriageway (two lanes in each direction) with a width of about 20 m and shoulder width of about 1–2 m on either side. There was no fencing along these highways, although vehicle guard rails were present in certain sections. The speed limit was 90 km/h on highway 185 and 110 km/h on highway 34. Average (±s.e.) vehicular traffic (counted at two different points on each highway between 0700 and 2300 h on six separate days) was 23 ± 5 cars, 16 ± 5 motorcycles and 10 ± 4 heavy vehicles per hour on highway 185; and 52 ± 11 cars, 24 ± 8 motorcycles and 20 ± 6 heavy vehicles per hour on highway 34. Many wildlife crossings, primarily eco-viaducts, have been constructed across Malaysia to restore connectivity between highway-bisected forest fragments, including forests in Kenyir and Sungai Yu. Eco-viaducts are bridge-like elevated roads (and highways) that allow passageway beneath for wildlife to safely cross between forests on either side of the highway (Figure 1d,e). Three eco-viaducts have been constructed across highway 185 in Kenyir and another three across highway 34 in Sungai Yu. The Kenyir eco-viaducts measuring 245 m, 140 m and 245 m in length were completed in 2008 (Figure 1a), while the Sungai Yu eco-viaducts measuring 80 m, 300 m and 1000 m were completed in 2014 (Figure 1b).

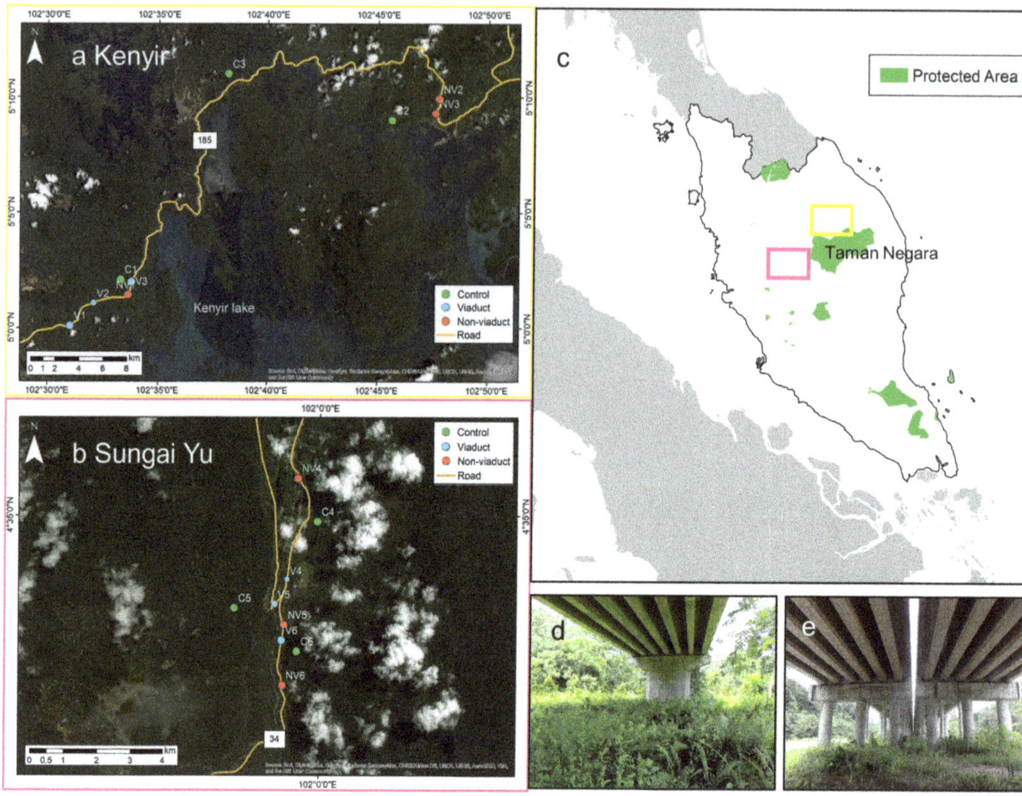

Figure 1. Sampling sites in (**a**) Kenyir, (**b**) Sungai Yu. Sites situated away from highways, along highways and adjacent to eco-viaducts are named "Control", "Non-viaduct" and "Viaduct", respectively. (**c**) Locations of study areas in Peninsular Malaysia (yellow: Kenyir; pink: Sungai Yu). Pictures of eco-viaducts (wildlife crossings) at (**d**) Kenyir and (**e**) Sungai Yu (photos by Tabitha Hui).

2.2. Small Mammal Trapping

Eighteen sites, six of them being non-viaduct highway-side sites, another six viaduct sites and six control sites (>500 m from any roads and highways), were selected for this study (Figure 1a,b). Both Kenyir and Sungai Yu had a set of nine sites each (Figure 1a,b). Each site was characterized by a pair of grids with 20 traps, each grid within a pair on opposite sides of the highway. The 20 traps within a grid were composed of ten Elliott sheet metal traps (32 × 10 × 10 cm) and ten Tomahawk wire cage traps (48 × 15 × 15 cm) (Figure S1) (except the 80 m wide viaduct in Sungai Yu which had only ten traps as there was not enough width across the viaduct to place 20 traps 10 m apart). Alternating Elliott and cage traps were set in parallel grids, with 10 m between traps (Figure 2). Paired grids were 50 m apart from each other, separated by the highway. The distance between grids was roughly chosen to be identical to the distance required for a crossing of the highway and its verges and ditches to confirm whether the highway accounts for any inhibition of movement. To avoid pseudoreplication issues, pairs of grids were at least 500 m apart, which is more than the home range of the small mammals in our study [45]. Trapping was conducted four times at Kenyir in 2017 and four times at Sungai Yu in 2018 from March to November, during the drier inter-monsoon and southwest monsoon seasons. Traps were set, checked and rebaited with bananas and peanut butter (cage traps) and vanilla scented oats (Elliott traps) [46] during five consecutive mornings and evenings to assess the diversity and abundance of both nocturnal and diurnal mammals. Captured individuals were identified to species, sex, age class and reproductive condition, weighed, measured and ear tagged before being released at the trap site. We collected ear clips for DNA tissue sampling from all animals trapped and tagged during live trapping using a 2 mm ear punch. Tissue samples were stored in absolute ethanol.

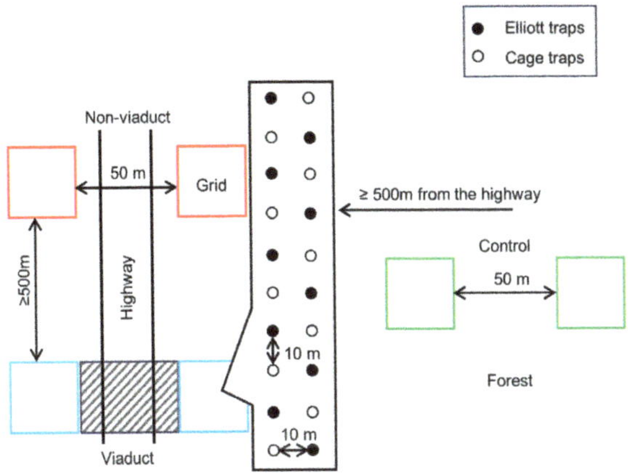

Figure 2. Schematic diagram of sampling design. The expanded grid box shows the layout of traps. Blue boxes: viaduct sites, red boxes: non-viaduct sites, green boxes: control sites.

In total, we trapped 448 individuals from 17 species. The species Whitehead's spiny rat *Maxomys whiteheadi* (Figure 3a), Rajah's spiny rat *Maxomys rajah* (Figure 3b), dark-tailed tree rat *Niviventer cremoriventer* (Figure 3c) and common tree shrew *Tupaia glis* (Figure 3d) were included in this study to capture a wide variety of biological and ecological characteristics. These four small mammal species are found in tall lowland forests and forest edge, and feed on insects and plant matter such as fruits and seeds. *M. rajah* and *T. glis* are similar in size and weight (100–200 g), while *M. whiteheadi* (35–80 g) and *N. cremoriventer* (50–100 g) are smaller. *M. whiteheadi* and *M. rajah* are both nocturnal ground dwelling rats; *N. cremoriventer* is a nocturnal tree rat which is a good climber and lives both arboreally and on the ground, *T. glis* is a diurnal treeshrew active on the ground and in the understory. Samples from 32 individuals (7 in Kenyir, 25 in Sungai Yu) of *M. rajah*, 39 (16 in Kenyir, 23 in Sungai Yu) of *M. whiteheadi*, 22 (20 in Kenyir, 2 in Sungai Yu) of *N. cremoriventer* and 45 (25 in Kenyir, 20 in Sungai Yu) of *T. glis* were selected based on their occurrences in pairs of sites (Table S1).

Figure 3. The four study species: (**a**) *Maxomys whiteheadi*, (**b**) *Maxomys rajah*, (**c**) *Niviventer cremoriventer* and (**d**) *Tupaia glis*. Photos by Tabitha Hui.

2.3. DNA Extraction and ddRAD-Seq Library Preparation

DNA extractions were performed using the DNeasy Blood & Tissue Kit (QIAGEN, Hilden, Germany) following the manufacturer's protocol for tissue. We prepared two libraries following Ng et al.'s [47] double-digest restriction-associated DNA sequencing (ddRADseq) protocol using EcoRI and MspI. To select for 250–600 bp fragments, as well as for the clean-up steps, we used Sera-Mag magnetic beads (Thermo Scientific, Waltham, MA, USA). DNA quantifications were performed with a Qubit 2.0 High Sensitivity DNA Assay (Invitrogen, Waltham, MA, USA). Before pooling samples, we checked DNA fragment size using a Fragment Analyser (Advanced Analytical Technologies, Inc., Ames, IA, USA). The two libraries were then spiked with 5% PhiX to prevent low nucleotide diversity issues from affecting the quality of the data, and were subsequently sequenced on an Illumina HiSeq 4000 platform (150 bp paired-end run).

Reads were demultiplexed and trimmed to 145 bp with *process_radtags* in STACKS 1.42 [48]. Reads with one or more uncalled bases were removed. For reference-based identification of ddRAD loci, we first aligned the demultiplexed reads of *T. glis* to the closely related *Tupaia chinensis* (GCA_000334495.1) [49], using the Burrows-Wheeler Aligner (BWA) [50] to index this reference genome. We used samtools 1.3.1 to convert the *sam* files to *bam* files, sort the aligned reads according to coordinates and filter files with a minimum required mapping quality score (MAPQ) score of 20 [51]. To call and filter single nucleotide polymorphisms (SNPs), we used *ref_map.pl* and *population* in STACKS 1.42 [48] for *T. glis*. In *population*, we set stack depth to 10 and the percentage of individuals represented at each locus to 0.9 and admitted only one random SNP per locus to preclude analysis of linked SNPs. For the other three species, there was no suitable reference genome, so SNPs were called de novo. In *ustacks*, we set the maximum distance (in nucleotides) allowed between stacks to 2, minimum depth of coverage required to create a stack to 3 and maximum distance allowed to align secondary reads to primary stacks to 4. We checked for SNPs under selection using BayeScan 2.1 [52] and used Plink 1.90 to remove linked loci and to calculate the level of missing data [53]. We allowed a variety of filters (0% or 10% missingness, including or excluding linked loci of $r^2 > 0.5$, and including or excluding minor allele frequency < 5%) to generate eight datasets for the preliminary testing to rule out potential sampling artefacts (e.g., non-random distribution of genotypes) (Table S2).

2.4. Statistical Analyses

To visualise genetic differentiation amongst individuals and identify potential population subdivision, we performed principal component analysis (PCA) using the R package SNPRelate 1.6.6 [54] for all eight datasets (four species at two study areas). We carried out sensitivity analysis, checking across various settings of missingness, linkage and minor allele frequencies by confirming the consistency of PCAs across eight datasets for all four species. As a consequence of this sensitivity analysis, we selected the dataset with 10% missingness, in which we excluded linked loci but included minor alleles for all subsequent analyses (Table S2). To understand the genetic differentiation between the two study areas, we calculated the Weir-Cockerham's F_{ST} between the two study areas using VCFtools v4.1 [55] and individual-pairwise relatedness using maximum likelihood estimation as implemented in SNPRelate.

We calculated pairwise genetic distances between all individuals within a species with the R package *poppr* [56]. To evaluate the efficacy of viaducts in facilitating the dispersal of small mammals and test the barrier effects of highways, Kruskal–Wallis tests were conducted on the genetic distances of individuals between pairs of grids within a site for each species across the three treatment types (viaduct, non-viaduct and control) and two study areas. This means that every individual of one species on one side of the highway or pair (in the case of control sites) is compared with every other individual of the same species on the other side of the highway or pair in the same pair of grids (site).

We also compared pairwise genetic distances among all individuals within a species in relation to different spatial distances to determine how resistance to dispersal changes in different landscape types. Using the Least Cost Path function in ArcGIS 10.6.1, we modeled three types of spatial distances, namely Euclidean distance (for examining the effects of isolation by distance (IBD)), least cost distance considering roads (and highways) as agents of resistance (not considering the potential effects of viaducts), and least cost distance considering roads as agents of resistance and with viaducts facilitating movement. The resistance value for non-road areas and viaducts was set at 0 to represent low-cost distance, and for roads and water bodies set at 1 to represent the high cost of crossing roads and water bodies in the least-cost distance calculations. We calculated the correlations between genetic and spatial distances for each species at each treatment grid using Mantel tests with 99 permutations (default value) and Monte-Carlo corrected p values, using the R package *ade4* [57]. One-tailed tests were used as the IBD model predicts that genetic differentiation will be positively correlated with increasing geographic distance [58]. All statistical analyses were conducted in R version 4.0.2 (R Core Team, 2020). Significance was considered at the $\alpha = 0.05$ level.

3. Results

We obtained ~9000 SNPs using STACKS for the complete data set consisting of all four species and retained ~5000 SNPs, after removing SNPs under selection and disequilibrium (Table S2). For each species, PCAs across different datasets were generally consistent, therefore, we chose the SNP dataset filtering 10% missingness and absolute linkage (>0.95) for all subsequent analyses.

PCA did not reveal any clear patterns of clustering in genetic variation amongst viaduct, non-viaduct and control sites (Figure 4, Figures S2 and S3). However, PCA revealed slight genomic differentiation in *M. whiteheadi* between study areas along principal component 1 (Figure 5). All four species exhibited low genetic differentiation (F_{ST}) between the two study areas (*M. whiteheadi* $F_{ST} = 0.022113378$, *M. rajah* $F_{ST} = 0.055480442$, *N. cremoriventer* $F_{ST} = 0.069197229$, *T. glis* $F_{ST} = 0.015547675$). We found related individuals (relatedness > 0.1) between the two study areas in *M. rajah* and *T. glis*, suggesting occasional exchange of individuals between the two study areas.

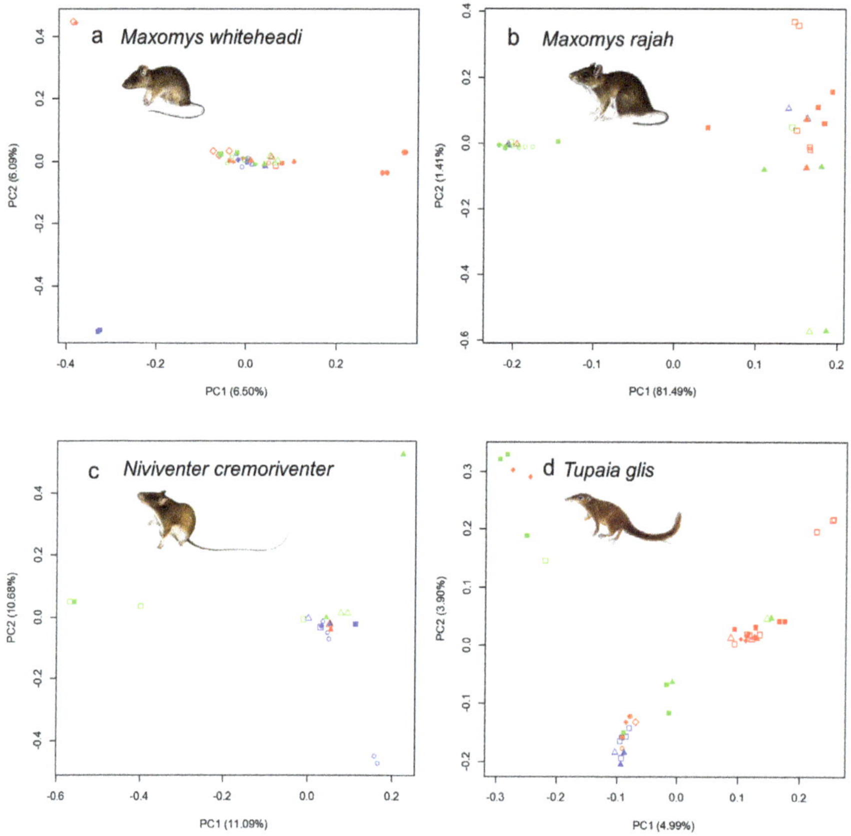

Figure 4. Principal component analysis (axes PC1 and PC2) of genetic differentiation among sampled individuals. (**a**) *Maxomys whiteheadi*, (**b**) *Maxomys rajah*, (**c**) *Niviventer cremoriventer* and (**d**) *Tupaia glis*. Blue: viaduct sites, red: non-viaduct sites, green: control sites. Same shape: same pair, filled/open shapes: opposite sides of the road of the same pair.

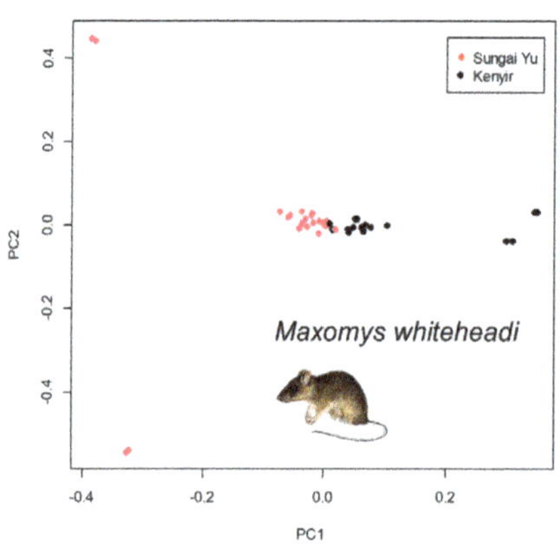

Figure 5. Principal component analysis (axes PC1 and PC2) of genetic differentiation among sampled individuals of *Maxomys whiteheadi* showing slight separation between Sungai Yu and Kenyir populations.

Kruskal–Wallis tests revealed that there were differences in the ranking of treatment types by genetic distance for *M. whiteheadi* ($\chi^2_{2,79}$ = 28.22, $p < 0.0001$, Figure 6a) and *M. rajah* ($\chi^2_{2,87}$ = 39.01, $p < 0.0001$, Figure 6b). Genetic distances between individuals were lowest at viaduct sites for *M. whiteheadi*. For *M. rajah*, genetic distances were highest at viaduct sites and lowest at non-viaduct sites, with control sites in between. There were no significant differences in genetic distances between treatment types for *N. cremoriventer* ($\chi^2_{2,27}$ = 5.38, $p = 0.068$, Figure 6c) and *T. glis* ($\chi^2_{2,101}$ = 3.39, $p = 0.18$, Figure 6d).

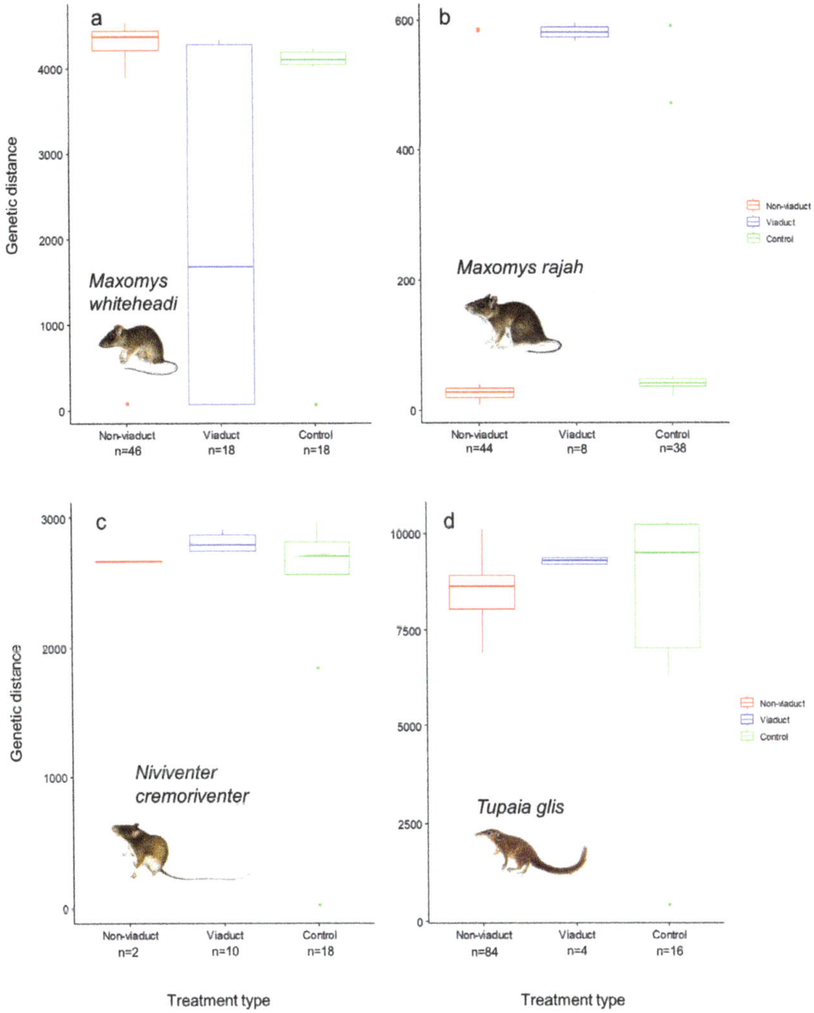

Figure 6. Comparisons of genetic distances between non-viaduct, viaduct and control sites. (**a**) *Maxomys whiteheadi*, (**b**) *Maxomys rajah*, (**c**) *Niviventer cremoriventer* and (**d**) *Tupaia glis*. The boxes, lines in the middle, whiskers and dots represent the interquartile range, median, minimum and maximum values (no more than 1.5 times the interquartile range) and outliers, respectively.

Study area was also a significant factor influencing genetic distances between individuals within a site for *M. whiteheadi* ($\chi^2_{1,79}$ = 7.18, $p < 0.01$, Figure 7a), with individuals from Sungai Yu showing more genetic differentiation than at Kenyir. Likewise, *T. glis* also showed significantly more genetic differentiation at Sungai Yu ($\chi^2_{1,101}$ = 37.35, $p < 0.0001$,

Figure 7d). *M. rajah* ($\chi^2_{1,87} = 0.72$, $p = 0.40$, Figure 7b) did not show differences in genetic distances between study areas. Due to low sample sizes at Sungai Yu, only *N. cremoriventer* samples from Kenyir were included in the analysis (Figure 7c).

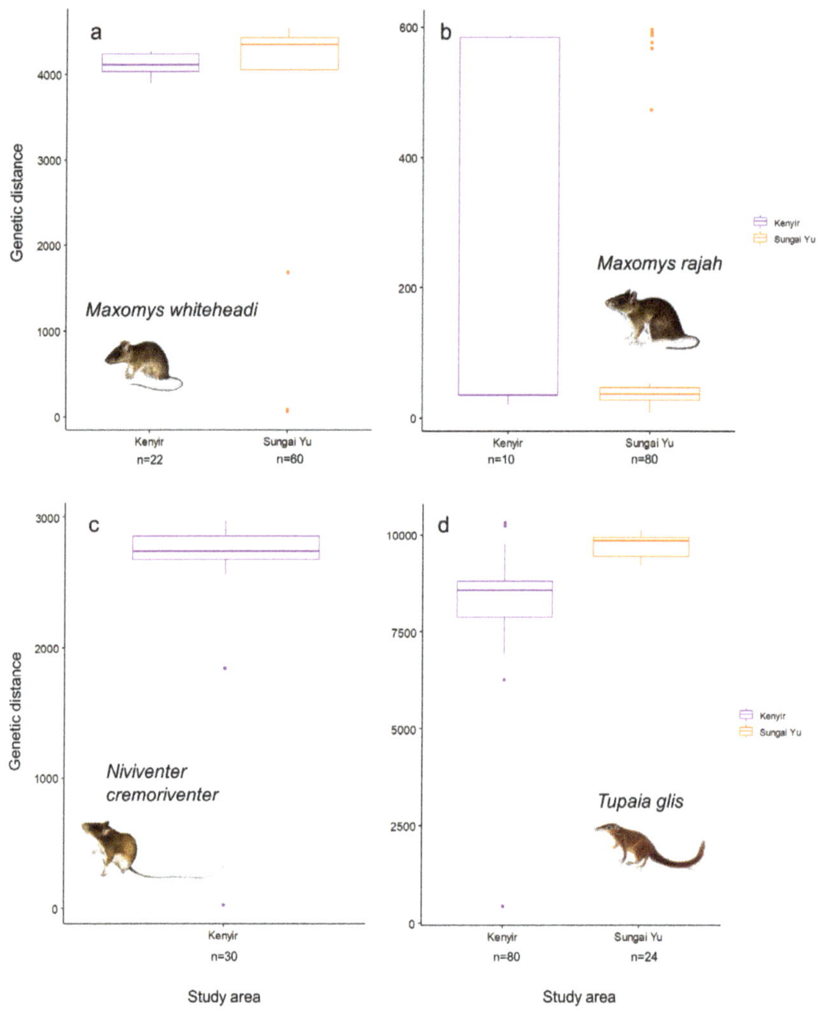

Figure 7. Comparisons of genetic distances between study areas. (**a**) *Maxomys whiteheadi*, (**b**) *Maxomys rajah*, (**c**) *Niviventer cremoriventer* and (**d**) *Tupaia glis*. The boxes, lines in the middle, whiskers and dots represent the interquartile range, median, minimum and maximum values (no more than 1.5 times the interquartile range) and outliers, respectively.

The dispersal of *M. whiteheadi* was found to be most consistent with isolation by distance (IBD) (highest correlation with Euclidean distance), indicating genetic and spatial distances between individuals were significantly positively correlated in both Kenyir and Sungai Yu, with minimal barrier effects of roads or facilitation effects of viaducts (Table 1). In Sungai Yu, the dispersal of *M. rajah* and *T. glis* correlated best with least cost distance, considering roads as resistance and viaducts as corridors facilitating movement, showing that they apparently prefer using viaducts to cross road barriers. No significant correlation with spatial distances were found in Kenyir for *M. rajah*, *T. glis* and *N. cremoriventer*. The paired samples of *N. cremoriventer* in Sungai Yu were insufficient for comparisons (Table 1).

Table 1. Mantel's r correlations between the genetic distances of each species and each spatial distance model (Euclidean: Euclidean distance; Road: least cost distance considering roads; Viaduct: least cost distance considering roads as resistance and viaducts facilitating movement). Results in bold are the spatial distance models with the highest correlation for the species in each study area, where multiple spatial distance models were significant.

		Kenyir		Sungai Yu	
	Spatial Distance Model	Mantel *r*	Simulated *p*-Value	Mantel *r*	Simulated *p*-Value
Maxomys rajah	Euclidean	0.118	0.15	0.355	0.01
	Road	0.117	0.16	0.373	0.01
	Viaduct	0.117	0.14	**0.506**	0.01
Maxomys whiteheadi	Euclidean	**0.245**	0.01	0.082	0.02
	Road	0.237	0.01	0.124	0.19
	Viaduct	0.237	0.01	0.126	0.22
Niviventer cremoriventer	Euclidean	0.096	0.1		
	Road	0.112	0.12		
	Viaduct	0.117	0.1		
Tupaia glis	Euclidean	0.121	0.17	0.113	0.17
	Road	0.102	0.19	0.204	0.09
	Viaduct	0.101	0.23	**0.207**	0.05

4. Discussion

Our study showed that eco-viaducts may facilitate movement and the maintenance of gene flow in a landscape bisected by roads at least in some species such as *M. whiteheadi*. Presumably, the higher genetic similarity of individuals at viaduct sites in these species, relative to that at non-viaduct sites, is due to movement of individuals through the viaducts [59], thus reducing population subdivision caused by habitat fragmentation.

Highways did not seem to pose a barrier to gene flow in *M. rajah* (Figure 6b). Surprisingly, the greatest genetic differentiation amongst the three treatment types for *M. rajah* was at viaduct sites. All spatial distance model correlations were significant as well, indicating that viaducts were not essential for *M. rajah* to cross roads. This differential impact of roads is likely because for a less agile ground dwelling small mammal such as *M. rajah*, the effort taken to cross natural structures in forests such as dense vegetation and uneven ground, or a longer route to reach a viaduct and go through it, could be similar to crossing artificial structures such as roads (Figure 6b). Perhaps *M. rajah* only uses the viaducts for dispersal to other more favourable habitats, as viaducts may also facilitate the movements of its predators such as civets and leopard cats [60].

Movement and gene flow in *M. whiteheadi* were influenced by road barriers as well as the study area. The *M. whiteheadi* population in Kenyir was slightly differentiated from the population in Sungai Yu. Their dispersal was also most aligned with the Euclidean distance model in both study areas, pointing to IBD. A distinction between populations was not observed in the other three species. A lack of differentiation between the two study areas for the other three species is not surprising as the study areas are relatively near, there are no severe breaks in forest cover at a landscape scale, and genetic differentiation is not only the result of isolation by distance. This is why we chose to use individual-based spatial analyses to investigate the subtle differentiation.

Greater genetic connectivity was observed for both *M. whiteheadi* and *T. glis* at Kenyir than at Sungai Yu. The highway at Sungai Yu is more than double the width of the highway at Kenyir and it is probably more of a barrier to movement than the narrower Kenyir highway. Thus, the significance of viaducts in facilitating movement and increasing genetic connectivity across wider roads such as at Sungai Yu would be greater than at Kenyir. This was probably the case for *M. whiteheadi* as it also showed the lowest genetic differentiation in viaduct sites. Captures of *M. whiteheadi* individuals within the viaducts themselves suggest that *M. whiteheadi* does use viaducts to cross roads and may even have part of their home ranges overlapping the viaducts. Of interest was the observation that for *T. glis*, individuals at viaduct sites were still genetically more dissimilar than those at control sites. This shows that despite the addition of viaducts to connect fragmented populations, they will not be able to restore genetic connectivity to match that of the original intact forest.

In Kenyir, the dispersal of *M. rajah*, *N. cremoriventer* and *T. glis* did not correlate significantly with any of the least cost distance models (Table 1), suggesting that the dispersal ability of these three species goes beyond the spatial scale of the study in this area. The highway bisecting the forest in Kenyir is much narrower than the highway in Sungai Yu, likely presenting itself as less of a barrier to movement. *M. rajah* has been found as road kills in Kenyir, evidence that it does cross roads directly if narrow enough [59]. *T. glis* is a relatively mobile species, able to disperse >4 km [61]; one individual in this study was trapped in two viaducts, a distance separation of >5 km.

There was no clear pattern of clustering in genetic variation amongst viaduct, non-viaduct and control sites in intact forest (Figure 4, Figures S2 and S3), suggesting that there are no significant differences in genetic composition of small mammal species amongst the three habitats. However, there was slight genetic differentiation between the *M. whiteheadi* populations at Kenyir and Sungai Yu (Figure 5). As the smallest species in this study, it is expected to have the least dispersal ability and may be starting to show genetic divergence between populations separated by greater distances.

The upgrading of the highways and construction of the viaducts in this study were relatively recent; therefore, population genetic data may not fully reflect the evolutionary change if the generation time is long. It is possible that not enough time has passed to allow the vegetation in the viaducts to grow and establish stable corridors or habitats. Mills and Allendorf [62] suggested that only one migrant per generation is sufficient to maintain genetic diversity while allowing some divergence between populations. Rosenberg et al. [24] also argue that corridors are more effective at maintaining movement between populations for habitat specialists rather than for habitat generalists. All four species in this study are tolerant to disturbed forests and forest edge, and their diets are varied, consisting of insects, fruit and seeds. This may explain why the viaducts did not increase genetic connectivity as much as expected.

Our study shows that viaducts can increase genetic connectivity, but are not effective across all species and roads or highways. The effectiveness of the viaducts in reducing the barrier effects of roads depends on the target species' dispersal abilities, generation time, perception of landscape characteristics, road width, viaduct age and vegetation structural maturity. Viaducts work better when they have a structure which is more conducive for movement than a direct crossing across the road. However, from this study we observed that viaducts were not able to restore genetic connectivity to match that of the original intact forest and should therefore only be used as a last resort if it is not possible to keep the forest intact.

5. Conclusions

Our study shows that wildlife crossings such as viaducts may assist in maintaining gene flow within populations of certain small mammals in areas with relatively wider roads. In a managed landscape, there must be a balance between the economic benefits of building roads and the ecological benefits to species in maintaining connectivity. We have shown that one of the benefits of constructing and maintaining viaducts is an increase in gene flow within populations in some species, which may result in an increase in population persistence and a decrease in inbreeding.

Supplementary Materials: The following supporting information can be downloaded at: https://www.mdpi.com/article/10.3390/ani14030426/s1, Figure S1: Photos of trap types used in this study; Figure S2: Principal component analysis of genetic differentiation among sampled individuals (PC3 and PC4); Figure S3: Principal component analysis of genetic differentiation among sampled individuals (PC5 and PC6); Table S1: Individual species samples and sites at which they were captured; Table S2: Number of SNPs at each stage of filtering.

Author Contributions: Conceptualization, T.C.Y.H., Q.T. and F.E.R.; methodology, T.C.Y.H., Q.T., E.Y.X.N. and F.E.R.; validation, Q.T., E.Y.X.N., E.M.S. and F.E.R.; formal analysis, T.C.Y.H., Q.T., E.Y.X.N. and F.E.R.; investigation, T.C.Y.H., Q.T., E.Y.X.N., J.L.C. and F.E.R.; resources, F.E.R.; data curation, T.C.Y.H. and J.L.C.; writing—original draft preparation, T.C.Y.H.; writing—review and editing, Q.T., E.Y.X.N., J.L.C., E.M.S. and F.E.R.; visualization, T.C.Y.H., Q.T., E.Y.X.N. and F.E.R.; supervision, E.M.S. and F.E.R.; project administration, J.L.C., Q.T. and F.E.R.; funding acquisition, F.E.R. All authors have read and agreed to the published version of the manuscript.

Funding: Our research was financially supported by the Rufford Foundation Grant for Nature Conservation, grant number 21698-1, an IDEA WILD Equipment Grant, Skyrail Rainforest Foundation Research Grant and Animal Behavior Society Student Research Grant. Labwork and analytical work for this project was funded by a Singapore Ministry of Education (MOE) Tier 2 grant to FER (WBS number R-154-000-C41-112).

Institutional Review Board Statement: We would like to thank the Economic Planning Unit (Permit number UPE 40/200/19/3363) of the Prime Minister's Department and the Department of Wildlife and National Parks (HQ, Terengganu and Pahang) (Permit numbers JPHL&TN(IP):100-34/1.24 JLD.8, JPHL&TN(IP):100-34/1.24 JLD.12, JPHL&TN(IP):90-7/8 JLD.16 and NRE 600-2/2/21 JLD.6) for granting permission to undertake this project, as well as the Forestry Department of Peninsular Malaysia (Terengganu and Pahang) (Permit number JH/100 JLD.16) for granting permission to enter the forest reserves in Kenyir and Sungai Yu.

Informed Consent Statement: Not applicable.

Data Availability Statement: All available data are presented in this manuscript.

Acknowledgments: We would like to thank the staff and students of Universiti Malaysia Terengganu, Universiti Sains Malaysia, Department of Wildlife and National Parks (PERHILITAN), Rimba, Ecoteer and MYCAT staff and volunteers, Sohud bin Yaacob and Dome Nikong for providing the necessary facilities and assistance. We would also like to thank Tan Poai Ean and the late Dr Lim Boo Liat for assistance with identification of the specimens. We also acknowledge all respondents for their cooperation and knowledge sharing in this study as well as local guides that assisted us throughout this research.

Conflicts of Interest: The authors declare no conflicts of interest. The funders had no role in the design of the study; in the collection, analyses, or interpretation of data; in the writing of the manuscript; or in the decision to publish the results.

References

1. Fahrig, L.; Rytwinski, T. Effects of roads on animal abundance: An empirical review and synthesis. *Ecol. Soc.* **2009**, *14*, 1. [CrossRef]
2. Debinski, D.M.; Holt, R.D. A survey and overview of habitat fragmentation experiments. *Conserv. Biol.* **2000**, *14*, 342–355. [CrossRef]
3. Laurance, W.F. Contemporary drivers of habitat fragmentation. In *Global Forest Fragmentation*; CABI: Wallingford, UK, 2014; pp. 20–27.
4. Robert McDonald, W.; Cassady St. Clair, C. The effects of artificial and natural barriers on the movement of small mammals in Banff National Park, Canada. *Oikos* **2004**, *105*, 397–407. [CrossRef]
5. Cros, E.; Ng, E.Y.; Oh, R.R.; Tang, Q.; Benedick, S.; Edwards, D.P.; Tomassi, S.; Irestedt, M.; Ericson, P.G.; Rheindt, F.E. Fine-scale barriers to connectivity across a fragmented South-East Asian landscape in six songbird species. *Evol. Appl.* **2020**, *13*, 1026–1036. [CrossRef] [PubMed]
6. Rivera-Ortíz, F.; Aguilar, R.; Arizmendi, M.; Quesada, M.; Oyama, K. Habitat fragmentation and genetic variability of tetrapod populations. *Anim. Conserv.* **2015**, *18*, 249–258. [CrossRef]
7. Chattopadhyay, B.; Garg, K.M.; Mendenhall, I.H.; Rheindt, F.E. Historic DNA reveals Anthropocene threat to a tropical urban fruit bat. *Curr. Biol.* **2019**, *29*, R1299–R1300. [CrossRef] [PubMed]

8. Ewers, R.M.; Didham, R.K. Confounding factors in the detection of species responses to habitat fragmentation. *Biol. Rev.* **2006**, *81*, 117–142. [CrossRef]
9. Christie, M.R.; Knowles, L.L. Habitat corridors facilitate genetic resilience irrespective of species dispersal abilities or population sizes. *Evol. Appl.* **2015**, *8*, 454–463. [CrossRef] [PubMed]
10. Haddad, N.M.; Bowne, D.R.; Cunningham, A.; Danielson, B.J.; Levey, D.J.; Sargent, S.; Spira, T. Corridor use by diverse taxa. *Ecology* **2003**, *84*, 609–615. [CrossRef]
11. Noss, R.F. Corridors in real landscapes: A reply to Simberloff and Cox. *Conserv. Biol.* **1987**, *1*, 159–164. [CrossRef]
12. Bennett, A.F. Habitat corridors and the conservation of small mammals in a fragmented forest environment. *Landsc. Ecol.* **1990**, *4*, 109–122. [CrossRef]
13. Saunders, D.A.; Hobbs, R.J.; Margules, C.R. Biological consequences of ecosystem fragmentation: A review. *Conserv. Biol.* **1991**, *5*, 18–32. [CrossRef]
14. Mabry, K.E.; Barrett, G.W. Effects of corridors on home range sizes and interpatch movements of three small mammal species. *Landsc. Ecol.* **2002**, *17*, 629–636. [CrossRef]
15. Clevenger, A.P.; Chruszcz, B.; Gunson, K. Drainage culverts as habitat linkages and factors affecting passage by mammals. *J. Appl. Ecol.* **2001**, *38*, 1340–1349. [CrossRef]
16. McDonald, W.; St Clair, C.C. Elements that promote highway crossing structure use by small mammals in Banff National Park. *J. Appl. Ecol.* **2004**, *41*, 82–93. [CrossRef]
17. Bellis, M.A.; Griffin, C.R.; Warren, P.; Jackson, S.D. Utilizing a multi-technique, multi–taxa approach to monitoring wildlife passageways in southern vermont. *Oecologia Aust.* **2013**, *17*, 111–128. [CrossRef]
18. Beier, P.; Noss, R.F. Do habitat corridors provide connectivity? *Conserv. Biol.* **1998**, *12*, 1241–1252. [CrossRef]
19. Haddad, N.M. Finding the corridor more traveled. *Proc. Natl. Acad. Sci. USA* **2008**, *105*, 19569–19570. [CrossRef]
20. Gonzalez, A.; Lawton, J.; Gilbert, F.; Blackburn, T.; Evans-Freke, I. Metapopulation dynamics, abundance, and distribution in a microecosystem. *Science* **1998**, *281*, 2045–2047. [CrossRef]
21. Dunning, J.B.; Borgella, R.; Clements, K.; Meffe, G.K. Patch isolation, corridor effects, and colonization by a resident sparrow in a managed pine woodland. *Conserv. Biol.* **1995**, *9*, 542–550. [CrossRef]
22. Fahrig, L.; Merriam, G. Habitat patch connectivity and population survival. *Ecology* **1985**, *66*, 1762–1768. [CrossRef]
23. Haddad, N.M.; Baum, K.A. An experimental test of corridor effects on butterfly densities. *Ecol. Appl.* **1999**, *9*, 623–633. [CrossRef]
24. Rosenberg, D.K.; Noon, B.R.; Meslow, E.C. Biological corridors: Form, function, and efficacy. *BioScience* **1997**, *47*, 677–687. [CrossRef]
25. Haddad, N.M. Corridor and distance effects on interpatch movements: A landscape experiment with butterflies. *Ecol. Appl.* **1999**, *9*, 612–622. [CrossRef]
26. Tewksbury, J.J.; Levey, D.J.; Haddad, N.M.; Sargent, S.; Orrock, J.L.; Weldon, A.; Danielson, B.J.; Brinkerhoff, J.; Damschen, E.I.; Townsend, P. Corridors affect plants, animals, and their interactions in fragmented landscapes. *Proc. Natl. Acad. Sci. USA* **2002**, *99*, 12923–12926. [CrossRef] [PubMed]
27. Haas, C.A. Dispersal and use of corridors by birds in wooded patches on an agricultural landscape. *Conserv. Biol.* **1995**, *9*, 845–854. [CrossRef]
28. Mech, S.G.; Hallett, J.G. Evaluating the effectiveness of corridors: A genetic approach. *Conserv. Biol.* **2001**, *15*, 467–474. [CrossRef]
29. Sutcliffe, O.L.; Thomas, C.D. Open corridors appear to facilitate dispersal by ringlet butterflies (*Aphantopus hyperantus*) between woodland clearings. *Conserv. Biol.* **1996**, *10*, 1359–1365. [CrossRef]
30. Beier, P. Dispersal of juvenile cougars in fragmented habitat. *J. Wildl. Manag.* **1995**, *59*, 228–237. [CrossRef]
31. Aars, J.; Ims, R.A. The effect of habitat corridors on rates of transfer and interbreeding between vole demes. *Ecology* **1999**, *80*, 1648–1655. [CrossRef]
32. Hale, M.L.; Lurz, P.W.; Shirley, M.D.; Rushton, S.; Fuller, R.M.; Wolff, K. Impact of landscape management on the genetic structure of red squirrel populations. *Science* **2001**, *293*, 2246–2248. [CrossRef]
33. Orrock, J.L. Conservation corridors affect the fixation of novel alleles. *Conserv. Genet.* **2005**, *6*, 623–630. [CrossRef]
34. Orrock, J.L.; Damschen, E.I. Corridors cause differential seed predation. *Ecol. Appl.* **2005**, *15*, 793–798. [CrossRef]
35. Weldon, A.J.; Haddad, N.M. The effects of patch shape on Indigo Buntings: Evidence for an ecological trap. *Ecology* **2005**, *86*, 1422–1431. [CrossRef]
36. Kjeldsen, S.R.; Zenger, K.R.; Leigh, K.; Ellis, W.; Tobey, J.; Phalen, D.; Melzer, A.; FitzGibbon, S.; Raadsma, H.W. Genome-wide SNP loci reveal novel insights into koala (*Phascolarctos cinereus*) population variability across its range. *Conserv. Genet.* **2016**, *17*, 337–353. [CrossRef]
37. Szulkin, M.; Gagnaire, P.A.; Bierne, N.; Charmantier, A. Population genomic footprints of fine-scale differentiation between habitats in Mediterranean blue tits. *Mol. Ecol.* **2016**, *25*, 542–558. [CrossRef]
38. Angeloni, F.; Wagemaker, N.; Vergeer, P.; Ouborg, J. Genomic toolboxes for conservation biologists. *Evol. Appl.* **2012**, *5*, 130–143. [CrossRef] [PubMed]
39. Sawaya, M.A.; Kalinowski, S.T.; Clevenger, A.P. Genetic connectivity for two bear species at wildlife crossing structures in Banff National Park. *Proc. R. Soc. B Biol. Sci.* **2014**, *281*, 20131705. [CrossRef]
40. Braunisch, V.; Segelbacher, G.; Hirzel, A.H. Modelling functional landscape connectivity from genetic population structure: A new spatially explicit approach. *Mol. Ecol.* **2010**, *19*, 3664–3678. [CrossRef]

41. Burkart, S.; Gugerli, F.; Senn, J.; Kuehn, R.; Bolliger, J. Evaluating the functionality of expert-assessed wildlife corridors with genetic data from roe deer. *Basic Appl. Ecol.* **2016**, *17*, 52–60. [CrossRef]
42. Brunke, J.; Russo, I.-R.M.; Orozco-terWengel, P.; Zimmermann, E.; Bruford, M.W.; Goossens, B.; Radespiel, U. Dispersal and genetic structure in a tropical small mammal, the Bornean tree shrew (*Tupaia longipes*), in a fragmented landscape along the Kinabatangan River, Sabah, Malaysia. *BMC Genet.* **2020**, *21*, 1–13. [CrossRef]
43. Steele, G.M.; Randolph, S.E. An experimental evaluation of conventional control measures against the sheep tick, *Ixodes ricinus* (L.)(Acari: Ixodidae). I. A unimodal seasonal activity pattern. *Bull. Entomol. Res.* **1985**, *75*, 489–500. [CrossRef]
44. Hecht, L.; Allcock, M. Potential Effects of Habitat Fragmentation on Wild Animal Welfare. 2020. Available online: https://www.researchgate.net/publication/342292473_Potential_effects_of_habitat_fragmentation_on_wild_animal_welfare (accessed on 13 November 2023).
45. Francis, C.M.; Barrett, P. *A Guide to the Mammals of Southeast Asia*; New Holland Publishers (UK) Ltd.: London, UK, 2008.
46. Hoffmann, A.; Decher, J.; Rovero, F.; Schaer, J.; Voigt, C.; Wibbelt, G. Field methods and techniques for monitoring mammals. *Man. Field Rec. Tech. Protoc. All Taxa Biodivers. Invent.* **2010**, *8*, 482–529.
47. Ng, E.Y.; Li, S.; Zhang, D.; Garg, K.M.; Song, G.; Martinez, J.; Hung, L.M.; Tu, V.T.; Fuchs, J.; Dong, L. Genome-wide SNPs confirm plumage polymorphism and hybridisation within a Cyornis flycatcher species complex. *Zool. Scr.* **2023**, *52*, 1–16. [CrossRef]
48. Catchen, J.; Hohenlohe, P.A.; Bassham, S.; Amores, A.; Cresko, W.A. Stacks: An analysis tool set for population genomics. *Mol. Ecol.* **2013**, *22*, 3124–3140. [CrossRef]
49. Li, H.; Durbin, R. Fast and accurate short read alignment with Burrows–Wheeler transform. *Bioinformatics* **2009**, *25*, 1754–1760. [CrossRef]
50. Lerner, H.R.; Meyer, M.; James, H.F.; Hofreiter, M.; Fleischer, R.C. Multilocus resolution of phylogeny and timescale in the extant adaptive radiation of Hawaiian honeycreepers. *Curr. Biol.* **2011**, *21*, 1838–1844. [CrossRef]
51. Li, H.; Handsaker, B.; Wysoker, A.; Fennell, T.; Ruan, J.; Homer, N.; Marth, G.; Abecasis, G.; Durbin, R.; Subgroup, G.P.D.P. The sequence alignment/map format and SAMtools. *Bioinformatics* **2009**, *25*, 2078–2079. [CrossRef]
52. Foll, M.; Gaggiotti, O. A genome-scan method to identify selected loci appropriate for both dominant and codominant markers: A Bayesian perspective. *Genetics* **2008**, *180*, 977–993. [CrossRef] [PubMed]
53. Chang, C.C.; Chow, C.C.; Tellier, L.C.; Vattikuti, S.; Purcell, S.M.; Lee, J.J. Second-generation PLINK: Rising to the challenge of larger and richer datasets. *Gigascience* **2015**, *4*, s13742-13015-10047-13748. [CrossRef] [PubMed]
54. Zheng, X. SNPRelate: Parrallel computing toolset for genome-wide association studies. *R Package Version* **2012**, *95*, B9.
55. Danecek, P.; Auton, A.; Abecasis, G.; Albers, C.A.; Banks, E.; DePristo, M.A.; Handsaker, R.E.; Lunter, G.; Marth, G.T.; Sherry, S.T. The variant call format and VCFtools. *Bioinformatics* **2011**, *27*, 2156–2158. [CrossRef]
56. Kamvar, Z.N.; Tabima, J.F.; Grünwald, N.J. Poppr: An R package for genetic analysis of populations with clonal, partially clonal, and/or sexual reproduction. *PeerJ* **2014**, *2*, e281. [CrossRef]
57. Bougeard, S.; Dray, S. Supervised multiblock analysis in R with the ade4 package. *J. Stat. Softw.* **2018**, *86*, 1–17. [CrossRef]
58. Wright, S. Isolation by distance. *Genetics* **1943**, *28*, 114. [CrossRef] [PubMed]
59. Hui, T.C.; Slade, E.M.; Chong, J.L. Roadkills in Northern Peninsular Malaysia. *Front. Environ. Sci.* **2021**, *9*, 637462. [CrossRef]
60. Hedges, L.; Clements, G.; Aziz, S.; Yap, W.; Laurance, S.; Goosem, M.; Laurance, W. Small carnivore records from a threatened habitat linkage in Terengganu, Peninsular Malaysia. *Small Carniv. Conserv.* **2013**, *49*, 9–14.
61. Mariana, A.; Shukor, M.; Muhd, N.H.; Intan, N.B.; Ho, T. Movements and home range of a common species of tree–shrew, Tupaia glis, surrounding houses of otoacariasis cases in Kuantan, Pahang, Malaysia. *Asian Pac. J. Trop. Med.* **2010**, *3*, 427–434. [CrossRef]
62. Mills, L.S.; Allendorf, F.W. The one-migrant-per-generation rule in conservation and management. *Conserv. Biol.* **1996**, *10*, 1509–1518. [CrossRef]

animals

MDPI

Article

Comprehensive Gene Expression Profiling Analysis of Adipose Tissue in Male Individuals from Fat- and Thin-Tailed Sheep Breeds

Sana Farhadi [1], Karim Hasanpur [1,*], Jalil Shodja Ghias [1], Valiollah Palangi [2], Aristide Maggiolino [3,*] and Vincenzo Landi [3]

1 Department of Animal Science, Faculty of Agriculture, University of Tabriz, Tabriz 51666-16471, Iran; farhadiso16@tabrizu.ac.ir (S.F.); shodja@tabrizu.ac.ir (J.S.G.)
2 Department of Animal Science, Faculty of Agriculture, Ege University, 35100 Izmir, Türkiye; valiollah.palangi@ege.edu.tr
3 Department of Veterinary Medicine, University of Bari Aldo Moro, 70010 Valenzano, Italy; vincenzo.landi@uniba.it
* Correspondence: karimhasanpur@tabrizu.ac.ir (K.H.); aristide.maggiolino@uniba.it (A.M.)

Simple Summary: For this paper, we investigated the differences in adipose tissue deposition between sheep breeds with fat and thin tails, relying on advanced techniques like meta-analyses and machine learning to analyze gene expression data. Our findings revealed key genes associated with fat metabolism, shedding light on the genetic factors influencing tail fat in sheep. Notably, three specific genes (*POSTN*, *K35*, and *SETD4*) were identified as significant biosignatures related to fat deposition. This innovative approach (combining data analysis and machine learning) enhances our understanding of how to optimize fat deposition in sheep breeds, which holds potential for more efficient animal breeding strategies and carcass fat reduction.

Abstract: It has been shown that tail fat content varies significantly among sheep breeds and plays a significant role in meat quality. Recently, significant efforts have been made to understand the physiological, biochemical, and genomic regulation of fat deposition in sheep tails in order to unravel the mechanisms underlying energy storage and adipose tissue lipid metabolism. RNA-seq has enabled us to provide a high-resolution snapshot of differential gene expression between fat- and thin-tailed sheep breeds. Therefore, three RNA-seq datasets were meta-analyzed for the current work to elucidate the transcriptome profile differences between them. Specifically, we identified hub genes, performed gene ontology (GO) analysis, carried out enrichment analyses of the Kyoto Encyclopedia of Genes and Genomes (KEGG) pathways, and validated hub genes using machine learning algorithms. This approach revealed a total of 136 meta-genes, 39 of which were not significant in any of the individual studies, indicating the higher statistical power of the meta-analysis. Furthermore, the results derived from the use of machine learning revealed *POSTN*, *K35*, *SETD4*, *USP29*, *ANKRD37*, *RTN2*, *PRG4*, and *LRRC4C* as substantial genes that were assigned a higher weight (0.7) than other meta-genes. Among the decision tree models, the Random Forest ones surpassed the others in adipose tissue predictive power fat deposition in fat- and thin-tailed breeds (accuracy > 0.85%). In this regard, combining meta-analyses and machine learning approaches allowed for the identification of three important genes (*POSTN*, *K35*, *SETD4*) related to lipid metabolism, and our findings could help animal breeding strategies optimize fat-tailed breeds' tail sizes.

Keywords: fat deposition; fat-tailed sheep; machine learning; RNA-seq

1. Introduction

Sheep are the leading meat and wool producers [1,2], with 20–25% of their world population being fat-tailed [3,4]. These sheep were first recorded on an Uruk III stone vessel about 5000 years ago [5]. These breeds are used in the different lamb production systems that are currently adopted around the world, reflecting different breeders' economic conditions, consumers' preferences, resources, and production aims. However, traditionally, sheep breeding is chiefly based on dairy breeds for both milk and meat production [2], with lamb being considered a high-quality product and even a delicacy in many countries [6].

In several breeds, artificial and natural selection have indirectly led to the development of adaptation to varying environmental conditions in different geographic regions. Within this spectrum, fat-tailed sheep are a noteworthy category of the world sheep population [7]. These sheep are primarily found in the Middle East, North and East Africa, and Central Asia. As highlighted by Xu et al. [8], fat tails serve as vital energy reserves that are crucial for survival in the wake of challenging conditions like droughts and food scarcity. This notion was further affirmed by Mwacharo et al. [9], who underscored that fat-tailed sheep predominate in the deserts and highlands of northern Africa, as well as in the semi-arid and arid regions of eastern and southern Africa.

The level of lipid storage in the carcass influences meat quality [10,11]. Moreover, fat affects many physical and chemical properties (e.g., color, water holding capacity) that are fundamental in the purchasing decision process [12–15]. Also, considering the increase in human living standards, people prefer tasty and healthy meat. Hence, increasing attention has been paid to provide leaner meat and to produce meat with intermuscular fat characterized by a lower saturation and higher unsaturation of fatty acids [16]. Adipose tissue is an important storage location for excess energy [10], with tail and subcutaneous fat being domestic animals' major fat storage sites [11]. The number of sheep breeds that have evolved worldwide is very high, with many being found specifically in northern Africa, the Middle East, Central Asia, and Western China [17]. It is assumed that the first home sheep were thin, but over time, due to the need to store energy for harsh environmental conditions, fat-tailed breeds gradually appeared [18]. However, in modern sheep industry systems, thin-tailed breeds are more desirable, while tail tissue has lost its importance in fat-tailed sheep. There are several logical reasons behind this trend: 1. in modern sheep breeding systems, there is no need for energy from tail tissue because intensive or semi-intensive feeding systems are preferred; 2. feed efficiency is decreased due to the higher energy requirement of fat anabolism as compared to the generation of protein or other molecules 3. today, the health of consumers is threatened by the consumption of high-fat foods; 4. a large tail can cause problems for mating and animal welfare. Therefore, raising thin-tailed sheep is cost-effective for both producers and consumers, and one of the sheep industry's primary goals is to study lean meat. In this context, disentangling the molecular mechanism of fat accumulation is critical to reduce its content in the carcass, as the manipulation of fat deposition is crucial to produce lean meat.

To date, various genomic- [19–25] and transcriptomic-based studies [26–30] have aimed to pinpoint the wide range of genes responsible for fat deposition. Most studies have addressed the mechanism of fat deposition in the tail of fat-tailed sheep breeds [3,24,31–34], with the majority of them focusing on one gene, especially Leptin (*LEP*) [35–37], Fatty Acid Banding Protein4 (*FABP4*) [38–40], Adiponectin, C1Q And Collagen Domain Containing (*ADIPOQ*) [11,30], and Stearoyl-CoA Desaturase (*SCD*) [41,42]. Nowadays, instead of examining single genes, the whole genome of animals can be examined using next-generation sequencing (NGS) technologies. Using this approach, the whole transcriptomes of single cells can be examined using RNA sequencing (RNA-Seq). This method makes it possible to measure the expression of countless genes simultaneously and gives us a lot of information about the genome, even if there is little consensus on the obtained results, with the differentially expressed genes (DEGs) of one study not being supported by the results of another. The observed differences in the identified DEGs among studies requires a meta-analysis to uncover the genes that are responsible for fat deposition. Indeed, this method, through the use of rigorous statistical tests, can disclose patterns hidden in individual studies and allows one to draw conclusions with a high degree of reliability. For this study, by employing meta-analysis and machine learning approaches, we re-analyzed data from three recently conducted whole transcriptome RNA-seq studies of Guangling Large-Tailed and Small-Tailed Han sheep [11], Lori-Bakhtiari (fat-tailed) and Zel (thin-tailed) sheep [43], and Ghezel (fat-tailed) and Zel (thin-tailed) sheep [29]. The primary purpose of the current study was to identify differential meta-genes in male individuals of fat- and thin-tailed sheep breeds as transcriptomic signatures of fat deposition.

2. Materials and Methods

An overview of the process followed in this study is shown in Figure 1.

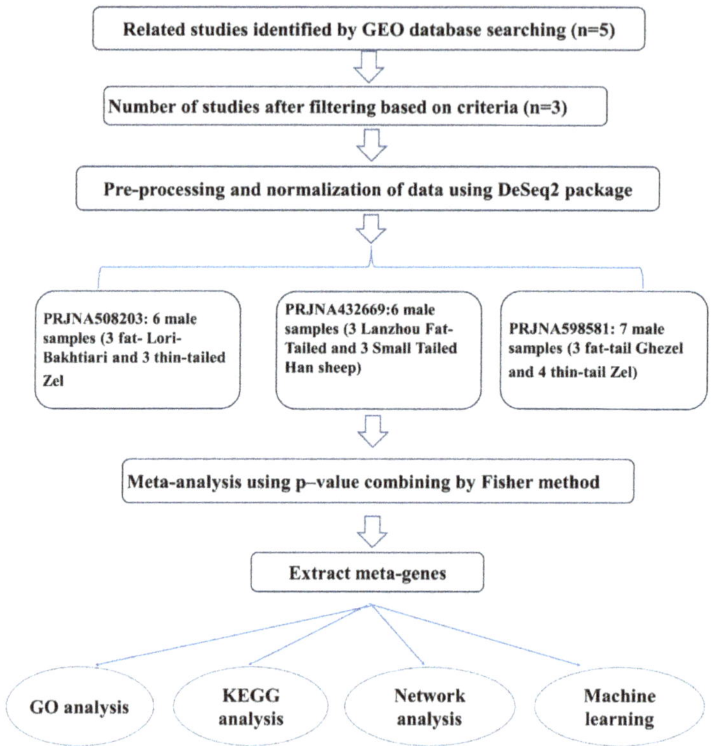

Figure 1. Flowchart of meta-analysis of the present study.

2.1. Dataset Collection

The keywords that were used were "Ovis aries", "Fat-tailed", "Thin-tailed", "Fat deposition" and "Lipid metabolism". We used PubMed Central (https://www.ncbi.nlm.nih.gov/pubmed accessed on 11 June 2021) and Google Scholar (https://scholar.google.com accessed on 11 June 2021). After identifying suitable RNA-seq studies of tail-fat deposition in the relevant fat- and thin-tailed sheep breeds, the related data were retrieved from either EMBL_EBI (https://www.ebi.ac.uk/arrayexpress accessed on 11 June 2021) or Gene Expression Omnibus (GEO) of NCBI (https://www.ncbi.nlm.nih.gov/gds accessed on 11 June 2021) databases. Two studies were excluded from our studies because the type of tissue examined or the type of RNA examined, or the sex of the samples were different. Finally, a set of sequencing data were collected from the fat tail tissue of male individuals from different sheep breeds.

2.2. Quality Control, Mapping, and Differential Gene Expression Analysis

Raw sequencing reads were subjected to quality control using FastQC (v0.11.5) [44] and trimmed using Trimmomatic software (v0.35); raw reads with adapter contamination and more than 10% of unknown bases, as well as with more than 50% of low-quality bases were trimmed out. Moreover, undesirable reads after trimming were filtered out [45]. The clean reads were mapped to the sheep reference genome v4.0 (ftp://ftp.ncbi.nlm.nih.gov/genomes/Ovis_aries/ accessed on 11 June 2021) using TopHat (v2.1.1) [46]. Sorted Binary Alignment Map (BAM) files were converted to *Sequence Alignment Map* (SAM) files, and count matrices were generated using htseq-count [47]. Then, the expression of the genes was normalized for library size and gene length to determine gene abundances using Fragment Per Kilo bases per Million (FPKM) [48], and the differentially expressed genes between the fat- and thin-tailed samples were identified using the DEseq2 package of R software [49]. For every dataset, each of these steps was carried out separately.

2.3. Meta-Analysis

The results of multiple scientific studies can be combined via a meta-analysis [50]. In addition to providing estimates of unknown effect sizes, meta-analyses can identify interesting and otherwise undetected relationships based on these results [51]. A set of p-values was computed for all three individual study datasets and later combined using the Fisher method of the meta RNAseq package [52]. Significance was set at $p < 0.05$. We named the final set of DEGs identified via our meta-analysis meta-genes.

2.4. GO Classification and KEGG Pathway Analysis

Gene ontology analysis is used to describe the function of genes in organisms. The Database for Annotation, Visualization, and Integrated Discovery (DAVID) (https://david.ncifcrf.gov/ accessed on 11 June 2021) was utilized to identify the category of meta-genes in the Gene Ontology (GO) based on Biological Processes (BP), Cellular Components (CC), and Molecular Functions (MF). Furthermore, the Kyoto Encyclopedia of Genes and Genomes (KEGG) pathway analysis tool (http://www.genome.jp/kegg/ accessed on 11 June 2021) was used to detect the metabolic pathways that are enriched by the meta-genes. For the enrichment analysis, terms with $p < 0.05$ generated using the modified Fisher Exact test were set as the cutoff thresholds.

2.5. Protein–Protein Network and Module Analysis

Protein functional interactions and their systematic properties help to provide context in molecular biology systems. The STRING database (http://string-db.org accessed on 11 June 2021) integrates protein–protein interactions that include direct (physical) and indirect (functional) interactions [53]. To predict protein–protein interactions, the identified meta-genes were imported into the STRING database (v11.0) [54]. The functional modules were detected via clustering using the K means algorithm. Also, the Cytoscape plugin cytoHubba (v3.7.2) was utilized to identify hub genes using the Maximal Clique Centrality (MCC) method [55]. The PPI networks were constructed based on co-expression, neighborhood interactions, text mining, gene fusion, and databases as interaction sources. Functional modules were defined in the constructed networks by clustering the K-means algorithm into three modules [56].

2.6. Validation of Hub Genes Using Machine Learning Algorithms

Machine learning (ML) is a subset of artificial intelligence (AI) that uses algorithms to automatically learn insights and identify patterns from data to make better decisions. Decision Tree (DT) is one of the simplest and best models in machine learning, the main purpose of which is to predict the value of the target variable by using learning simple decision rules deduced from data features [57]. To assess the effectiveness of hub genes in distinguishing between fat-tailed and thin-tailed sheep, meta-genes and their corresponding expression values were identified and subjected to gene selection using seven different weighting algorithms. Normalized data were used for the attribute weighting algorithms (AWs). A range between 0 and 1 was considered for all weights, with values closer to 1 indicating important attributes for one meta-gene. These algorithms include Uncertainty,

Relief, Gain Ratio, Information Gain, Gini Index, Chi-square, and Rule [58]. Only meta-genes with a weighting value greater than 0.7 were selected for DT construction using four criteria—Information Gain, Information Gain ratio, Gini index, and Accuracy—along with the leave-one-out cross-validation (LOOCV) method. During this process, the initial dataset was divided into training and testing sets. One sample at a time was removed from the initial dataset and added to the testing set, while all the others remained in the training set [59].

3. Results

3.1. Sequencing Data Collection

For this study, a total of approximately 200 Giga bases of RNA-seq from three datasets were utilized, comprising 19 samples in total. Each of the three datasets was sequenced using the Illumina HiSeq 2000 platform, and information pertaining to the three datasets is listed in Table 1.

Table 1. Summary information of three RNA-seq datasets sourced from sheep tail fat tissue.

GEO Accession Number	Number of Samples		Tissue Sample	Age of Slaughter (Month)	Read Length	Reference
	Thin-Tailed	Fat-Tailed				
PRJNA432669	3	3	Tail	6	150 bp	[11]
PRJNA508203	3	3	Tail	6	150 bp	[43]
PRJNA598581	4	3	Tail	6	150 bp	[29]

3.2. Meta-Analysis of RNA-Seq Data

A total of 136 meta-genes were identified. Fisher's method of differential analysis identified 20 meta-genes in PRJNA432669, 2 in PRJNA598581, and 75 in PRJA508203, along with 39 that had not been previously identified in the individual analyses.

3.3. Functional Enrichment Analysis of Meta-Genes

The top 10 BP terms are shown in Table 2.

Table 2. Top 10 Biological Process terms enriched by meta-genes.

Biological Process Terms	Adjusted *p*-Value
Positive regulation of T cell cytokine production	0.002
Extracellular matrix organization	0.004
Stress-activated MAPK cascade	0.008
ERK1 and ERK2 cascade	0.008
Positive regulation of interleukin-10 production	0.009
Positive regulation of interleukin-1 secretion	0.016
Positive regulation of interleukin-6 production	0.034
Positive regulation of interleukin-8 production	0.036
Positive regulation of interferon-gamma (IFN)secretion	0.04
Positive regulation of tumor necrosis factor (TNF) production	0.04

The CC terms of "lipid droplet", "Golgi lumen", "endoplasmic reticulum lumen", and MF terms of "lipoprotein lipase activity", "dipeptidyl-peptidase activity", "phospholipase activity", "interleukin-17 receptor activity", and "cAMP response element binding" were significantly enriched ($p < 0.05$). Several BP terms related to lipolysis, such as the "positive regulation of interleukin-1 beta secretion", "positive regulation of interleukin-6 production", "positive regulation of interleukin-8 production, "positive regulation of interleukin-

10 production", "regulation of interleukin-12 secretion", "regulation of interleukin-13 secretion", "positive regulation of interferon-gamma secretion", and "positive regulation of tumor necrosis factor production" were enriched by meta-genes. The meta-genes were also mapped onto the KEGG pathway database to identify the pathways related to fat deposition (Figure 2).

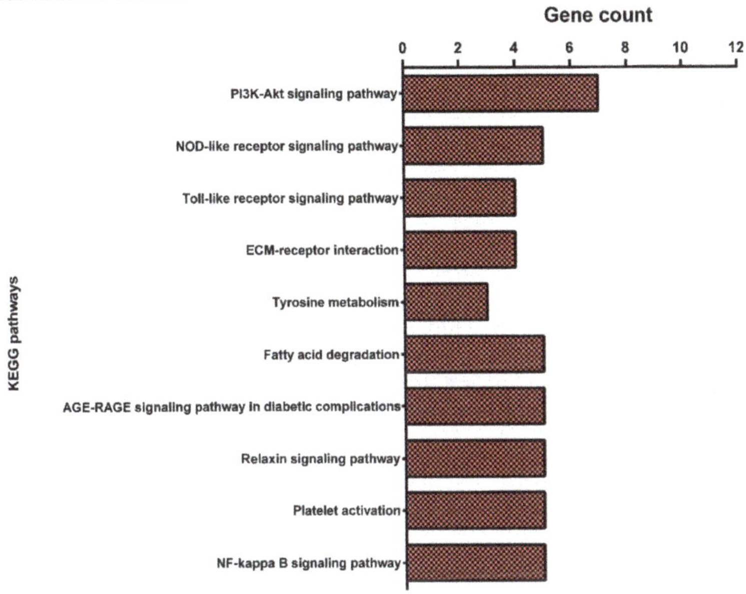

Figure 2. Top ten Kyoto Encyclopedia of Genes and Genomes (KEGG) pathways enriched by meta-genes.

3.4. Protein–Protein Interaction (PPI) Network and Module Analysis

The PPI network of meta-genes revealed that 88% of the identified meta-genes had considerable interaction with the primary functional modules based on the confidence score of the interaction (confidence score < 0.7). In contrast, other disconnected nodes had no interaction in PPI networks (Figure 3). Also, TNF Receptor Associated Factor 6 (*TRAF6*) and Collagen, type I, Alpha 1 (*COL1A1*) meta-genes were identified as hub genes in PPI networks' green and red modules, respectively.

3.5. Feature Selection for Machine Learning

The meta-analysis resulted in the identification of 136 differentially expressed genes between the fat- and thin-tailed sheep breeds. Ten meta-genes, including Periostin (*POSTN*), Keratin 35 (*K35*), SET Domain Containing 4 (*SETD4*), Ubiquitin Specific Peptidase 29 (*USP29*), Ankyrin Repeat Domain 37 (*ANKRD37*), *ENSOARG00000001454*, Reticulon (*RTN2*), Proteoglycan (*PRG4*), and Leucine Rich Repeat Containing 4C (*LRRC4C*), were detected by the majority of the attribute weighting algorithms (with weight above 0.7) as the most informative genes. The top ten meta-genes in the discrimination of fat- and thin-tailed samples, confirmed by the majority of AWs (with an average weight above 0.7), are reported in Table 3. These meta-genes gained higher importance than the remaining meta-genes and were believed to be more effective in distinguishing the two breeds. According to Figure 4, a mean expression comparison between two types of breeds was carried out using a two-sample *t*-test. The expression of *POSTN*, *K35*, and *SETD4* meta-genes showed significant differences between the two sheep breeds.

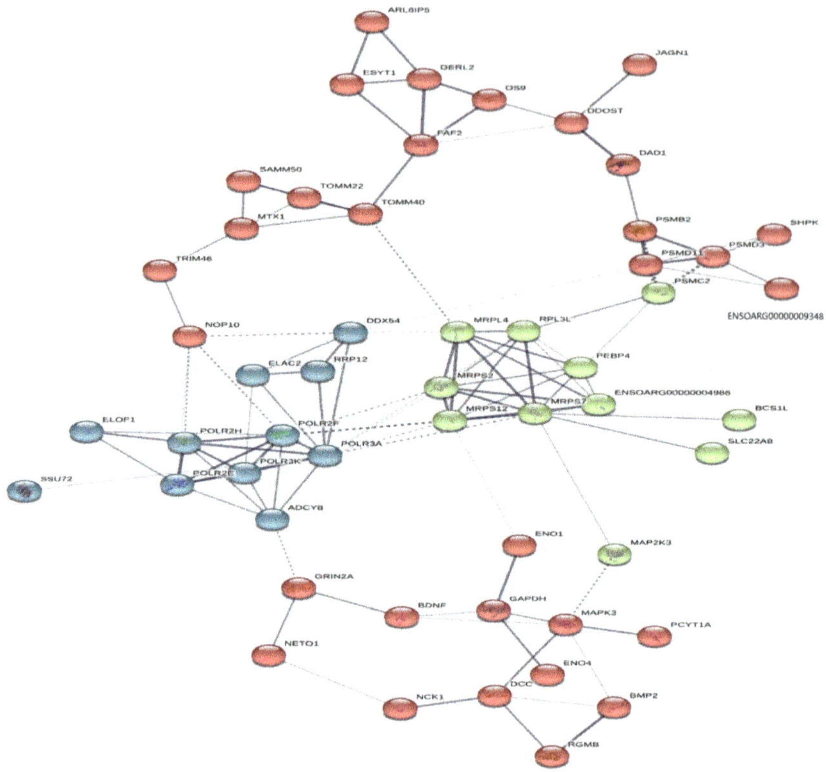

Figure 3. Protein–protein interaction (PPI) network and functional module analysis of meta-genes.

Table 3. Top 10 out of the 136 meta-genes according to seven attribute weighting algorithms (AWs).

Attribute	Weight_Info Gain Ratio	Weight_ Rule	Weight_Chi Squared	Weight_Gini Index	Weight_ Uncertainty	Weight_ Relief	Weight_Info Gain	Average_ Weight
POSTN	1	1	0.5	0.8	0.6	0.6	0.8	0.8
K35	0.8	0.9	0.6	0.6	0.7	1	0.6	0.8
SETD4	0.8	1.0	0.7	0.7	0.6	0.9	0.6	0.8
USP29	0.7	0.8	0.8	0.5	0.7	1.0	0.5	0.7
ANKRD37	0.7	0.9	0.8	0.5	0.7	0.9	0.5	0.7
ENSOARG 00000001454	0.8	1	0.5	0.6	0.4	0.7	0.6	0.7
RTN2	0.7	1	0.6	0.5	0.6	0.8	0.5	0.7
PRG4	0.8	1	0.3	0.6	0.4	0.9	0.6	0.7
LRRC4C	0.7	0.8	0.6	0.5	0.5	1	0.5	0.7

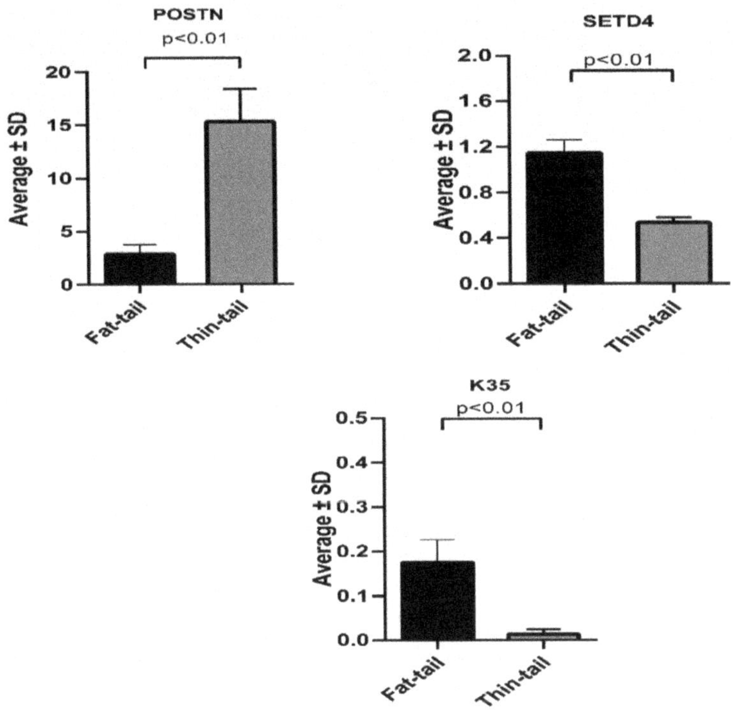

Figure 4. Three meta genes with attribute weighting above 0.8. A two-sample *t*-test was used for the mean comparisons.

The performances of the eight decision tree models are presented in Table 4. According to Table 4, among the decision tree models, the Random Forest with accuracy criterion and the Random Forest with gain_ratio criterion models surpassed the others in predicting fat deposition in both fat- and thin-tailed sheep breeds. These models had higher accuracy (above 0.85%).

Table 4. Performances of machine learning models in the distinction of fat- and thin-tailed sheep breeds via ten-fold cross validation.

Model	Accuracy
Random Forest with accuracy criterion	90% +/− 22.36%
Random Forest with gain_ratio criterion	85% +/− 13.69%
Decision Tree with gain_ratio criterion	58.33% +/− 37.27%
Decision Tree with accuracy criterion	75% +/− 35.36%
Deep Learning with Tanh parameter	85% +/− 22.36%
Deep Learning with Rectifier parameter	75% +/− 25.00%
Deep Learning with Maxout parameter	56.67% +/− 18.07%
Naïve Bayes	78.33% +/− 21.73%

Also, the machine learning results show that the two meta-genes with the highest weight (*K35* and *SETD4*) were down-regulated, and *POSTN* was up-regulated in the thin-tailed sheep breeds compared to the fat-tailed sheep breeds (Figure 5).

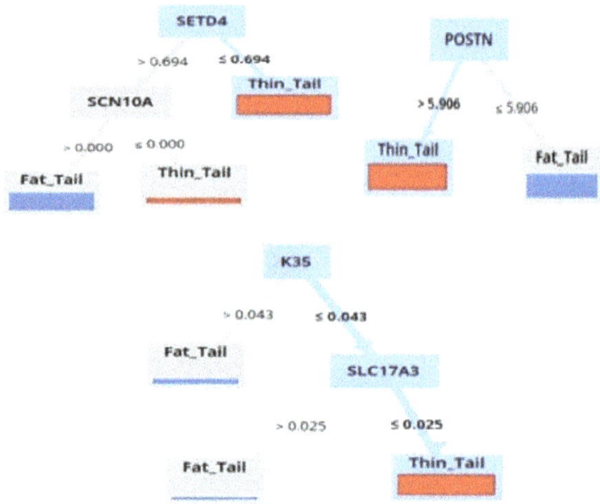

Figure 5. Decision tree induced by the Random Forest algorithm with gain_ratio criterion and accuracy criterion in distinguishing the fat-tailed sheep breeds from the thin-tailed sheep breeds in three meta-genes with attribute weighting above 0.8.

4. Discussion

The current study identified genes that are informative in terms of the tail fat deposition of fat-tailed sheep breeds through using, for the first time in the literature, a machine learning approach. "ERK1 and ERK2 cascade" and "stress-activated MAPK cascade" terms are related to lipid metabolism. The two extracellular signal-regulated kinases (ERKs), ERK1 and ERK2, are members of the mitogen-activated protein kinase (MAPK) pathway and participate in both cell differentiation and proliferation, as well as the regulation of lipolysis [60]. ERK activation leads to the fast stimulation of hormone-sensitive lipase (HSL) activity and contributes to increased lipolysis. Documented pieces of evidence show that, as a stimulator of lipolysis, catecholamines cannot only activate cAMP-dependent protein kinase (PKA) but also activate ERKs of the MAPK pathway [61]. Five genes were enriched in the "ERK1 and ERK2 cascade", including Kinase Insert Domain Receptor (*KDR*) or Vascular Endothelial Growth Factor Receptor 2 (*VEGFR2*), Galectin-9 (*LGALS9*), *TRAF6*, Nucleotide Binding Oligomerization Domain Containing 2 (*NOD2*), and Vascular Endothelial Growth Factor Receptor 3 (*VEGFR3*). Three of them (i.e., *KDR*, *NOD2*, and *LGALS9*) are closely related to lipid metabolism. A recent study showed that the *KDR* gene protects mice from obesity via fat burning and progressing lipolysis and enhancing basal metabolic rate [62]. In addition, it has been shown that galectin-9 enhances the production of microglial Tumor Necrosis Factor (TNF), which is the main lipolytic factor [63]. *NOD2* has also been shown to protect mice against diet-induced obesity and metabolic dysfunction, with obese mice lacking the *NOD2* gene suffering from metabolic dysfunction, including blood lipids, hyperglycemia, and steatosis, and the mass of adipose tissue and large fat droplets in liver cells increases [64]. These results were consistent with those of a recent study on the difference in adipose tissue metabolic pathways in fat- and thin-tailed sheep breeds [34].

There is a direct connection between lipid metabolism and terms such as Interleukin-1 (*IL-1*) [65], Interleukin-6 (*IL-6*) [29], Interleukin-8 (*IL-8*) [66], Interleukin-10 (*IL-10*) [67], Interleukin-12 (*IL-12*) [68]. The "positive regulation of tumor necrosis" factor production is also one of the BP terms closely related to lipolysis. It has been reported that, in human adipocytes, TNF-α stimulates lipolysis via the elevation of intracellular cAMP, MAPK, and extracellular signal-related kinase (ERK) [69,70]. All the inflammatory pathways and the lipolytic ERK/MAPK/TNF pathways enhance lipolysis. In other words, these pathways improve pro-inflammatory cytokine expression and increase lipolytic activity. Finally, these results suggest that some essential lipolytic pathways (e.g., "MAPK signaling pathway" and "TNF signaling pathway") and inflammatory pathways (e.g., "positive regulation of *IL-1* secretion", "positive regulation of *IL-6* production", "positive regulation of *IL-8* production", and "positive regulation of *IL-10* production") are active in thin-tailed sheep breeds. There have been reports that *IL-1* and interferon-gamma (IFN) stimulate lipolysis in cultured adipocytes [71]. In addition, inflammatory cytokine *IL-6* and *IL-8* mRNA expressions are involved in lipopolysaccharide-induced lipolysis in human adipocytes [67]. Another study introduced *IL-6* as a hub gene in the fat lipolysis of thin-tailed sheep breeds [29]. This gene is well known to be a lipolytic factor that stimulates fat lipolysis and fatty acid oxidation in humans [72,73], dairy cows [74], rats [75], mice [76,77], and sheep [29]. Thus, the up-regulation of the aforementioned genes in thin-tailed sheep might be closely related to lipolysis.

Our KEGG pathway analysis of the meta-genes revealed significant pathways (adjusted p-value < 0.05). Recent findings show that a set of pathways, such as lipid metabolism, extracellular matrix (ECM) remodeling, molecular transport, and inflammatory response, are enriched by a set of functional genes that maintain lipid homeostasis in response to extreme environments in tailed animals [34]. In the present study, some pathways, including "fatty acid degradation", "NF-kappa B signaling pathway", "NOD-like receptors", and "Toll-like receptors", were all related to fat metabolism as an inflammatory response. One of the significant terms in thin-tailed sheep breeds compared with fat-tailed sheep breeds is the "fatty acid degradation" pathway, the pathway known for the lipolysis of adipocytes. Another considerable term is "NF-kappa B signaling pathway". NF-κB is important for TNF-α-induced lipolysis of adipose tissue. Tumor necrosis factor-α (TNF-α) increases lipolysis in adipose tissue via the MAPK pathway. Several meta-genes, including Myeloid Differentiation Primary Response 88 (*MYD88*), TGF-Beta Activated Kinase 1 (MAP3K7) Binding Protein 2 (*TAB2*), Interleukin-1 Receptor-associated Kinase 1 (*IRAK1*), Phospholipase C Gamma 2 (*PLCG2*), and *TRAF6*, were found to be enriched in the "NF-kappa B signaling pathway", which is closely related to lipid metabolism.

"ECM-receptor interaction" is another significant pathway that is central to adipogenesis and fat tissue architecture [78]. Fat accumulation is an inflammatory condition related to increased extracellular matrix gene expression [34,79]. However, a direct connection between ECM gene expression and fat tissue inflammation has not been reported. In a recent study, transcriptome analysis of two broiler chickens showed that the extracellular matrix receptor interaction signaling pathway is crucial to chicken meat quality. This pathway might change intramuscular fat content, affecting broiler meat flavor [80]. In another study, comparative transcriptome analysis of three adipose tissues (i.e., subcutaneous, intramuscular, and omental adipose tissue) showed that the interactions between transmembrane receptors of fat cells and ECM components depend on depot-specific adipogenesis [81]. Cell adhesion receptors and ECM components interact with each other, creating a complex network. According to one study, cell surface receptors receive signals from the ECM that influence growth, survival, migration, differentiation, and proliferation in maintaining cell homeostasis [82]. All enriched meta-genes in this pathway are up-regulated in thin-tailed sheep breeds, possibly due to the interaction between ECM components. This result is in accordance with a recent study that showed that fat-tailed sheep are less responsive to seasonal changes in inflammation and fat cell size, ECM regeneration, and lipid metabolism, which indicates the improvement of homeostasis [34].

"NOD-like receptors" (NLR) and "Toll-like receptors" (TLR) are two pattern recognition receptors that have severe roles in the inflammation of adipocytes and the immune response [83]. Both of the mentioned KEGG pathways were significantly enriched in the current study. The activation of a sub-family of these receptors [84–86] has been shown to stimulate lipolysis from adipose tissue or adipocytes. Therefore, all these pathways maintain fat homeostasis in response to extreme environments in sheep breeds.

The meta-analysis results revealed 39 meta-genes that were insignificant in each of the individual studies, indicating the higher statistical power of the meta-analysis in the discovery of biosignatures. In addition, the results derived from using machine learning showed that three significant genes (*POSTN*, *K35*, and *SETD4*) gained higher weights (>0.8) than others, according to the AW algorithms. Interestingly, these meta-genes, along with other genes with a weight >0.8, are associated with lipid metabolism. The decision tree induced by the Random Forest Model shows that *POSTN*, *K35*, and *SETD4* meta-genes directly affect lipid metabolism. Interestingly, the *SETD4* gene is one of the 39 meta-genes that were insignificant in the individual studies. Currently, the role of the *K35* gene in fat metabolism has not been identified, but two other genes have been shown to be related to fat metabolism. There have been reports that the *SETD4* gene has considerable potential for tumorigenesis. It is thought that the *SETD4* gene has proliferation potential in fat cells [87]. Thus, lipogenesis might be associated with the up-regulation of the *SETD4* gene in fat-tailed sheep. Moreover, the loss of *POSTN* attenuates lipid metabolism in adipose tissue [88]. Among the decision tree models, both the Random Forest with accuracy criterion and the Random Forest with gain_ratio criterion models outperformed others in the prediction of fat deposition in sheep breeds. These high-performance models enabled us to detect the *POSTN*, *K35*, and *SETD4* meta-genes as biosignatures or biomarkers for fat metabolism. Therefore, the combination of meta-analysis and machine learning approaches employed in the current study improved the power of discovering informative genes that may aid the progress of animal breeding strategies to optimize tail fat in fat-tailed breeds.

5. Conclusions

Fat deposition is a complex trait that requires comprehensive research to be elucidated. However, the integration of machine learning and meta-analyses approaches, as carried out for the current work, may help to better understand the most critical causal genes that can be exploited as strong biomarkers of fat deposition; in our study, three meta-genes, namely, *POSTN*, *K35*, and *SETD4*, were identified as strong biosignatures of fat deposition. Our findings may provide a base for strategies to optimize fat deposition in the tail of fat-tailed breeds, thus decreasing the fat content of carcasses.

Author Contributions: Conceptualization, S.F., K.H. and J.S.G.; methodology, K.H. and J.S.G.; software, S.F., K.H. and J.S.G.; validation, V.L., A.M. and V.P.; formal analysis, S.F. and K.H.; investigation, S.F., K.H. and J.S.G.; resources, J.S.G. and S.F.; data curation, S.F. and J.S.G.; writing—original draft preparation, S.F., K.H. and J.S.G.; writing—review and editing, A.M., V.P. and V.L.; visualization, V.P. and A.M.; supervision, V.P. and V.L.; project administration, S.F. and J.S.G.; funding acquisition, A.M. and V.P. All authors have read and agreed to the published version of the manuscript.

Funding: This study was carried out within the Agritech National Research Center via a grant from the European Union Next-GenerationEU (PIANO NAZIONALE DI RIPRESA E RESILIENZA (PNRR)—MISSIONE 4 COMPONENTE 2, INVESTIMENTO 1.4—D.D. 1032 17/06/2022, CN00000022). This manuscript reflects only the authors' views and opinions; neither the European Union nor the European Commission can be considered responsible for them.

Institutional Review Board Statement: Not applicable.

Informed Consent Statement: Not applicable.

Data Availability Statement: All used data are publically available. The accession numbers have been reported in Table 1.

Acknowledgments: We are grateful to Mohammad Farhadian for his kind help. The authors express their sincere gratitude to Selim Esen for his valuable cooperation and technical assistance.

Conflicts of Interest: The authors declare no conflict of interest.

References

1. Maggiolino, A.; Bragaglio, A.; Salzano, A.; Rufrano, D.; Claps, S.; Sepe, L.; Damiano, S.; Ciarcia, R.; Dinardo, F.R.; Hopkins, D.L.; et al. Dietary supplementation of suckling lambs with anthocyanins: Effects on growth, carcass, oxidative and meat quality traits. *Anim. Feed Sci. Technol.* **2021**, *276*, 114925. [CrossRef]
2. De Palo, P.; Maggiolino, A.; Centoducati, P.; Calzaretti, G.; Ceci, E.; Tateo, A. An assessment of sire-breed effects on carcass and meat quality traits of lambs at the ages of 40 and 100 days from Comisana ewes crossed with Suffolk or Bergamasca rams. *J. Anim. Prod. Sci.* **2018**, *58*, 1794–1801. [CrossRef]
3. Mohapatra, A.; Shinde, A. Fat-tailed sheep-an important sheep genetic resource for meat production in tropical countries: An overview. *Indian. J. Small Rumin.* **2018**, *24*, 1–17. [CrossRef]
4. Zhou, G.; Wang, X.; Yuan, C.; Kang, D.; Xu, X.; Zhou, J.; Geng, R.; Yang, Y.; Yang, Z.; Chen, Y. Integrating miRNA and mRNA expression profiling uncovers miRNAs underlying fat deposition in sheep. *BioMed Res. Int.* **2017**, *2017*, 1857580. [CrossRef] [PubMed]
5. Ryder, M.L. *Sheep and Man*; Gerald Duckworth & Co., Ltd.: London, UK, 1983.
6. Vieira, C.; Fernández, A.M. Effect of ageing time on suckling lamb meat quality resulting from different carcass chilling regimes. *Meat Sci.* **2014**, *96*, 682–687. [CrossRef]
7. Moradi, M.H.; Nejati-Javaremi, A.; Moradi-Shahrbabak, M.; Dodds, K.G.; McEwan, J.C. Genomic scan of selective sweeps in thin and fat tail sheep breeds for identifying of candidate regions associated with fat deposition. *BMC Genet.* **2012**, *13*, 10. [CrossRef] [PubMed]
8. Xu, S.-S.; Ren, X.; Yang, G.-L.; Xie, X.-L.; Zhao, Y.-X.; Zhang, M.; Shen, Z.-Q.; Ren, Y.-L.; Gao, L.; Shen, M.; et al. Genome-wide association analysis identifies the genetic basis of fat deposition in the tails of sheep (*Ovis aries*). *Anim. Genet.* **2017**, *48*, 560–569. [CrossRef] [PubMed]
9. Mwacharo, J.M.; Kim, E.-S.; Elbeltagy, A.R.; Aboul-Naga, A.M.; Rischkowsky, B.A.; Rothschild, M.F. Genomic footprints of dryland stress adaptation in Egyptian fat-tail sheep and their divergence from East African and western Asia cohorts. *Sci. Rep.* **2017**, *7*, 17647. [CrossRef]
10. Braissant, O.; Foufelle, F.; Scotto, C.; Dauça, M.; Wahli, W. Differential expression of peroxisome proliferator-activated receptors (PPARs): Tissue distribution of PPAR-alpha,-beta, and-gamma in the adult rat. *Endocrinology* **1996**, *137*, 354–366. [CrossRef]
11. Li, B.; Qiao, L.; An, L.; Wang, W.; Liu, J.; Ren, Y.; Pan, Y.; Jing, J.; Liu, W. Transcriptome analysis of adipose tissues from two fat-tailed sheep breeds reveals key genes involved in fat deposition. *BMC Genom.* **2018**, *19*, 338. [CrossRef]
12. Quiñones, J.; Maggiolino, A.; Bravo, S.; Muñoz, E.; Lorenzo, J.M.; Cancino, D.; Díaz, R.; Saenz, C.; Sepúlveda, N.; De Palo, P. Effect of canola oil on meat quality and fatty acid profile of Araucano creole lambs during fattening period. *Anim. Feed Sci. Technol.* **2019**, *248*, 20–26. [CrossRef]
13. De Palo, P.; Tateo, A.; Maggiolino, A.; Marino, R.; Ceci, E.; Nisi, A.; Lorenzo, J.M. Martina Franca donkey meat quality: Influence of slaughter age and suckling technique. *Meat Sci.* **2017**, *134*, 128–134. [CrossRef] [PubMed]
14. Tateo, A.; Maggiolino, A.; Domínguez, R.; Lorenzo, J.M.; Dinardo, F.R.; Ceci, E.; Marino, R.; Della Malva, A.; Bragaglio, A.; De Palo, P. Volatile Organic Compounds, Oxidative and Sensory Patterns of Vacuum Aged Foal Meat. *Animals* **2020**, *10*, 1495. [CrossRef] [PubMed]
15. Serrano, M.P.; Maggiolino, A.; Lorenzo, J.M.; De Palo, P.; García, A.; Landete-Castillejos, T.; Gambín, P.; Cappelli, J.; Domínguez, R.; Pérez-Barbería, F.J.; et al. Meat quality of farmed red deer fed a balanced diet: Effects of supplementation with copper bolus on different muscles. *Animal* **2019**, *13*, 888–896. [CrossRef]
16. Dong, L.; Jin, Y.; Cui, H.; Yu, L.; Luo, Y.; Wang, S.; Wang, H. Effects of diet supplementation with rumen-protected betaine on carcass characteristics and fat deposition in growing lambs. *Meat Sci.* **2020**, *166*, 108154. [CrossRef] [PubMed]
17. Pourlis, A.F. A review of morphological characteristics relating to the production and reproduction of fat-tailed sheep breeds. *Trop. Anim. Health Prod.* **2011**, *43*, 1267–1287. [CrossRef]
18. Kalds, P.; Luo, Q.; Sun, K.; Zhou, S.; Chen, Y.; Wang, X. Trends towards revealing the genetic architecture of sheep tail patterning: Promising genes and investigatory pathways. *Anim. Genet.* **2021**, *52*, 799–812. [CrossRef]
19. Ornaghi, M.G.; Guerrero, A.; Vital, A.C.P.; de Souza, K.A.; Passetti, R.A.C.; Mottin, C.; de Araújo Castilho, R.; Sañudo, C.; do Prado, I.N. Improvements in the quality of meat from beef cattle fed natural additives. *Meat Sci.* **2020**, *163*, 108059. [CrossRef]
20. Luo, R.; Zhang, X.; Wang, L.; Zhang, L.; Li, G.; Zheng, Z. GLIS1, a potential candidate gene affect fat deposition in sheep tail. *Mol. Biol. Rep.* **2021**, *48*, 4925–4931. [CrossRef]
21. Shao, J.; He, S.; Pan, X.; Yang, Z.; Nanaei, H.A.; Chen, L.; Li, R.; Wang, Y.; Gao, S.; Xu, H. Allele-specific expression reveals the phenotypic differences between thin-and fat-tailed sheep. *J. Genet. Genom.* **2020**, *49*, 583–586.
22. Moioli, B.; Pilla, F.; Ciani, E. Signatures of selection identify loci associated with fat tail in sheep. *J. Anim. Sci.* **2015**, *93*, 4660–4669. [CrossRef] [PubMed]
23. Wei, C.; Wang, H.; Liu, G.; Wu, M.; Cao, J.; Liu, Z.; Liu, R.; Zhao, F.; Zhang, L.; Lu, J. Genome-wide analysis reveals population structure and selection in Chinese indigenous sheep breeds. *BMC Genom.* **2015**, *16*, 194. [CrossRef]
24. Dong, K.; Yang, M.; Han, J.; Ma, Q.; Han, J.; Song, Z.; Luosang, C.; Gorkhali, N.A.; Yang, B.; He, X. Genomic analysis of worldwide sheep breeds reveals PDGFD as a major target of fat-tail selection in sheep. *BMC Genom.* **2020**, *21*, 800. [CrossRef] [PubMed]

25. Pan, Z.; Li, S.; Liu, Q.; Wang, Z.; Zhou, Z.; Di, R.; An, X.; Miao, B.; Wang, X.; Hu, W. Rapid evolution of a retro-transposable hotspot of ovine genome underlies the alteration of BMP2 expression and development of fat tails. *BMC Genom.* **2019**, *20*, 261. [CrossRef] [PubMed]
26. Wang, X.; Zhou, G.; Xu, X.; Geng, R.; Zhou, J.; Yang, Y.; Yang, Z.; Chen, Y. Transcriptome profile analysis of adipose tissues from fat and short-tailed sheep. *Gene* **2014**, *549*, 252–257. [CrossRef]
27. Ma, L.; Li, Z.; Cai, Y.; Xu, H.; Yang, R.; Lan, X. Genetic variants in fat-and short-tailed sheep from high-throughput RNA-sequencing data. *Anim. Genet.* **2018**, *49*, 483–487. [CrossRef]
28. Bakhtiarizadeh, M.R.; Alamouti, A.A. RNA-Seq based genetic variant discovery provides new insights into controlling fat deposition in the tail of sheep. *Sci. Rep.* **2020**, *10*, 13525. [CrossRef]
29. Farhadi, S.; Shodja Ghias, J.; Hasanpur, K.; Mohammadi, S.A.; Ebrahimie, E. Molecular mechanisms of fat deposition: IL-6 is a hub gene in fat lipolysis, comparing thin-tailed with fat-tailed sheep breeds. *Arch. Anim. Breed.* **2021**, *64*, 53–68. [CrossRef]
30. Zhang, W.; Xu, M.; Wang, J.; Wang, S.; Wang, X.; Yang, J.; Gao, L.; Gan, S. Comparative transcriptome analysis of key genes and pathways activated in response to fat deposition in two sheep breeds with distinct tail phenotype. *Front. Genet.* **2021**, *12*, 639030. [CrossRef]
31. Ibrahim, A.; Baliarti, E.; Budisatria, I.; Artama, W.T.; Widayanti, R.; Maharani, D.; Tavares, L.; Margawati, E.T. Genetic diversity and relationship among Indonesian local sheep breeds on Java Island based on mitochondrial cytochrome b gene sequences. *J. Genet. Eng. Biotechnol.* **2023**, *21*, 34. [CrossRef]
32. Deribe, B.; Beyene, D.; Dagne, K.; Getachew, T.; Gizaw, S.; Abebe, A. Morphological diversity of northeastern fat-tailed and northwestern thin-tailed indigenous sheep breeds of Ethiopia. *Heliyon* **2021**, *7*, e07472. [CrossRef]
33. Yazdani, H.; Gholizadeh, M.; Farhadi, A.; Moradi, M.H. Study of the copy number variation on the sex chromosome in some Iranian sheep breeds. *Iran. J. Anim. Sci.* **2023**, *54*, 253–266.
34. Xu, Y.-X.; Wang, B.; Jing, J.-N.; Ma, R.; Luo, Y.-H.; Li, X.; Yan, Z.; Liu, Y.-J.; Gao, L.; Ren, Y.-L. Whole-body adipose tissue multi-omic analyses in sheep reveal molecular mechanisms underlying local adaptation to extreme environments. *Commun. Biol.* **2023**, *6*, 159. [CrossRef] [PubMed]
35. Mohammadabadi, M.; Kord, M.; Nazari, M. Studying expression of leptin gene in different tissues of Kermani Sheep using Real Time PCR. *Agric. Biotechnol. J.* **2018**, *10*, 111–123.
36. Krawczyńska, A.; Herman, A.P.; Antushevich, H.; Bochenek, J.; Wojtulewicz, K.; Zieba, D.A. The Effect of Leptin on the Blood Hormonal Profile (Cortisol, Insulin, Thyroid Hormones) of the Ewe in Acute Inflammation in Two Different Photoperiodical Conditions. *Int. J. Mol. Sci.* **2022**, *23*, 8109. [CrossRef]
37. Picó, C.; Palou, M.; Pomar, C.A.; Rodríguez, A.M.; Palou, A. Leptin as a key regulator of the adipose organ. *Rev. Endocr. Metab. Disord.* **2022**, *23*, 13–30. [CrossRef]
38. Zhang, X.; Liu, C.; Kong, Y.; Li, F.; Yue, X. Effects of intramuscular fat on meat quality and its regulation mechanism in Tan sheep. *Front. Nutr.* **2022**, *9*, 908355. [CrossRef]
39. Ji, K.; Jiao, D.; Yang, G.; Degen, A.A.; Zhou, J.; Liu, H.; Wang, W.; Cong, H. Transcriptome analysis revealed potential genes involved in thermogenesis in muscle tissue in cold-exposed lambs. *Front. Genet.* **2022**, *13*, 1017458. [CrossRef]
40. Zhang, Y.-M.; Erdene, K.; Zhao, Y.-B.; Li, C.-Q.; Wang, L.; Tian, F.; Ao, C.-J.; Jin, H. Role of white adipose tissue browning in cold seasonal acclimation in grazing Mongolian sheep (*Ovis aries*). *J. Therm. Biol.* **2022**, *109*, 103333. [CrossRef]
41. Aali, M.; Moradi-Shahrbabak, H.; Moradi-Shahrbabak, M.; Sadeghi, M.; Kohram, H. Polymorphism in the SCD gene is associated with meat quality and fatty acid composition in Iranian fat-and thin-tailed sheep breeds. *Livest. Sci.* **2016**, *188*, 81–90. [CrossRef]
42. Aali, M.; Shahrbabak, M.M.; Shahrbabak, H.M.; Sadeghi, M. Identifying novel SNPs and allelic sequences of the stearoyl-Coa desaturase gene (*SCD*) in fat-tailed and thin-tailed sheep breeds. *Biochem. Genet.* **2014**, *52*, 153–158. [CrossRef]
43. Bakhtiarizadeh, M.R.; Salehi, A.; Alamouti, A.A.; Abdollahi-Arpanahi, R.; Salami, S.A. Deep transcriptome analysis using RNA-Seq suggests novel insights into molecular aspects of fat-tail metabolism in sheep. *Sci. Rep.* **2019**, *9*, 9203. [CrossRef]
44. Andrews, S. FastQC: A quality control tool for high throughput sequence data. *Anal. Biochem.* **2010**, *548*, 38–43.
45. Bolger, A.M.; Lohse, M.; Usadel, B. Trimmomatic: A flexible trimmer for Illumina sequence data. *Bioinformatics* **2014**, *30*, 2114–2120. [CrossRef] [PubMed]
46. Trapnell, C.; Pachter, L.; Salzberg, S.L. TopHat: Discovering splice junctions with RNA-Seq. *Bioinformatics* **2009**, *25*, 1105–1111. [CrossRef] [PubMed]
47. Anders, S.; Pyl, P.T. Wolfgang Huber HTSeq—A Python framework to work with high-throughput sequencing data. *Bioinformatics* **2015**, *31*, 166–169. [CrossRef]
48. Mortazavi, A.; Williams, B.A.; McCue, K.; Schaeffer, L.; Wold, B. Mapping and quantifying mammalian transcriptomes by RNA-Seq. *Nat. Methods* **2008**, *5*, 621. [CrossRef] [PubMed]
49. Love, M.; Anders, S.; Huber, W. Differential analysis of count data–the DESeq2 package. *Genome Biol.* **2014**, *15*, 10–1186.
50. Adèr, H.J.; Mellenbergh, G.J. *Research Methodology in the Social, Behavioural and Life Sciences: Designs, Models and Methods*; Sage: London, UK, 1999.
51. Conn, V.S.; Ruppar, T.M.; Phillips, L.J.; Chase, J.-A.D. Using meta-analyses for comparative effectiveness research. *Nurs. Outlook* **2012**, *60*, 182–190. [CrossRef]
52. Rau, A.; Marot, G.; Jaffrézic, F. Differential meta-analysis of RNA-seq data from multiple studies. *BMC Bioinform.* **2014**, *15*, 91. [CrossRef]

53. Szklarczyk, D.; Franceschini, A.; Wyder, S.; Forslund, K.; Heller, D.; Huerta-Cepas, J.; Simonovic, M.; Roth, A.; Santos, A.; Tsafou, K.P. STRING v10: Protein–protein interaction networks, integrated over the tree of life. *Nucleic Acids Res.* **2015**, *43*, D447–D452. [CrossRef]
54. Szklarczyk, D.; Morris, J.H.; Cook, H.; Kuhn, M.; Wyder, S.; Simonovic, M.; Santos, A.; Doncheva, N.T.; Roth, A.; Bork, P. The STRING database in 2017: Quality-controlled protein–protein association networks, made broadly accessible. *Nucleic Acids Res.* **2016**, *45*, 362–368. [CrossRef]
55. Chin, C.-H.; Chen, S.-H.; Wu, H.-H.; Ho, C.-W.; Ko, M.-T.; Lin, C.-Y. cytoHubba: Identifying hub objects and sub-networks from complex interactome. *BMC Syst. Biol.* **2014**, *8*, S11. [CrossRef]
56. Le, V.-H.; Kim, S.-R. K-strings algorithm, a new approach based on Kmeans. In Proceedings of the 2015 Conference on Research in Adaptive and Convergent Systems, New York, NY, USA, 9–12 October 2015; pp. 15–20.
57. Vabalas, A.; Gowen, E.; Poliakoff, E.; Casson, A.J. Machine learning algorithm validation with a limited sample size. *PLoS ONE* **2019**, *14*, e0224365. [CrossRef] [PubMed]
58. Ebrahimie, E.; Ebrahimi, M.; Sarvestani, N.R.; Ebrahimi, M. Protein attributes contribute to halo-stability, bioinformatics approach. *Saline Syst.* **2011**, *7*, 1. [CrossRef]
59. Farhadian, M.; Rafat, S.A.; Panahi, B.; Mayack, C. Weighted gene co-expression network analysis identifies modules and functionally enriched pathways in the lactation process. *Sci. Rep.* **2021**, *11*, 2367.
60. Ikoma-Seki, K.; Nakamura, K.; Morishita, S.; Ono, T.; Sugiyama, K.; Nishino, H.; Hirano, H.; Murakoshi, M. Role of LRP1 and ERK and cAMP signaling pathways in lactoferrin-induced lipolysis in mature rat adipocytes. *PLoS ONE* **2015**, *10*, e0141378. [CrossRef] [PubMed]
61. Carmen, G.-Y.; Víctor, S.-M. Signalling mechanisms regulating lipolysis. *Cell. Signal.* **2006**, *18*, 401–408. [CrossRef]
62. Robciuc, M.R.; Kivelä, R.; Williams, I.M.; de Boer, J.F.; van Dijk, T.H.; Elamaa, H.; Tigistu-Sahle, F.; Molotkov, D.; Leppänen, V.-M.; Käkelä, R. VEGFB/VEGFR1-induced expansion of adipose vasculature counteracts obesity and related metabolic complications. *Cell Metab.* **2016**, *23*, 712–724. [CrossRef]
63. Steelman, A.J.; Li, J. Astrocyte galectin-9 potentiates microglial TNF secretion. *J. Neuroinflammation* **2014**, *11*, 144. [CrossRef]
64. Rodriguez-Nunez, I.; Caluag, T.; Kirby, K.; Rudick, C.N.; Dziarski, R.; Gupta, D. Nod2 and Nod2-regulated microbiota protect BALB/c mice from diet-induced obesity and metabolic dysfunction. *Sci. Rep.* **2017**, *7*, 548. [CrossRef]
65. Matsuki, T.; Horai, R.; Sudo, K.; Iwakura, Y. IL-1 plays an important role in lipid metabolism by regulating insulin levels under physiological conditions. *J. Exp. Med.* **2003**, *198*, 877–888. [CrossRef]
66. Bruun, J.M.; Pedersen, S.B.; Richelsen, B. Regulation of interleukin 8 production and gene expression in human adipose tissue in vitro. *J. Clin. Endocrinol. Metab.* **2001**, *86*, 1267–1273. [CrossRef]
67. Acosta, J.R.; Tavira, B.; Douagi, I.; Kulyte, A.; Arner, P.; Ryden, M.; Laurencikiene, J. Human-specific function of IL-10 in adipose tissue linked to insulin resistance. *J. Clin. Endocrinol. Metab.* **2019**, *104*, 4552–4562. [CrossRef]
68. Sari, A.; Dogan, S.; Nibali, L.; Koseoglu, S. Evaluation of IL-23p19/Ebi3 (IL-39) gingival crevicular fluid levels in periodontal health, gingivitis, and periodontitis. *Clin. Oral. Investig.* **2022**, *26*, 7209–7218. [CrossRef]
69. Jurga, L.; Elisabet, A.N.m.; Andrea, D.; Lennart, B.; Erik, N.s.; Dominique, L.; Peter, A.; Mikael, R.n. NF-κB is important for TNF-α-induced lipolysis in human adipocytes. *J. Lipid Res.* **2007**, *48*, 1069–1077. [CrossRef]
70. Zhang, H.H.; Halbleib, M.; Ahmad, F.; Manganiello, V.C.; Greenberg, A.S. Tumor necrosis factor-α stimulates lipolysis in differentiated human adipocytes through activation of extracellular signal-related kinase and elevation of intracellular cAMP. *Diabetes* **2002**, *51*, 2929–2935. [CrossRef]
71. Feingold, K.R.; Doerrler, W.; Dinarello, C.A.; Fiers, W.; Grunfeld, C. Stimulation of lipolysis in cultured fat cells by tumor necrosis factor, interleukin-1, and the interferons is blocked by inhibition of prostaglandin synthesis. *Endocrinology* **1992**, *130*, 10–16. [CrossRef] [PubMed]
72. Van Hall, G.; Steensberg, A.; Sacchetti, M.; Fischer, C.; Keller, C.; Schjerling, P.; Hiscock, N.; Møller, K.; Saltin, B.; Febbraio, M.A. Interleukin-6 stimulates lipolysis and fat oxidation in humans. *J. Clin. Endocrinol. Metab.* **2003**, *88*, 3005–3010. [CrossRef] [PubMed]
73. Xu, Y.; Zhang, Y.; Ye, J. IL-6: A potential role in cardiac metabolic homeostasis. *Int. J. Mol. Sci.* **2018**, *19*, 2474. [CrossRef] [PubMed]
74. Contreras, G.A.; Strieder-Barboza, C.; Raphael, W. Adipose tissue lipolysis and remodeling during the transition period of dairy cows. *J. Anim. Sci. Biotechnol.* **2017**, *8*, 41. [CrossRef]
75. Nonogaki, K.; Fuller, G.M.; Fuentes, N.L.; Moser, A.H.; Staprans, I.; Grunfeld, C.; Feingold, K.R. Interleukin-6 stimulates hepatic triglyceride secretion in rats. *Endocrinology* **1995**, *136*, 2143–2149. [CrossRef]
76. Ma, Y.; Gao, M.; Sun, H.; Liu, D. Interleukin-6 gene transfer reverses body weight gain and fatty liver in obese mice. *Biochim. Biophys. Acta* **2015**, *1852*, 1001–1011. [CrossRef]
77. Han, J.; Meng, Q.; Shen, L.; Wu, G. Interleukin-6 induces fat loss in cancer cachexia by promoting white adipose tissue lipolysis and browning. *Lipids Health Dis.* **2018**, *17*, 14. [CrossRef]
78. Mariman, E.C.; Wang, P. Adipocyte extracellular matrix composition, dynamics and role in obesity. *Cell. Mol. Life Sci.* **2010**, *67*, 1277–1292. [CrossRef]
79. Adapala, V.J.; Adedokun, S.A.; Considine, R.V.; Ajuwon, K.M. Acute inflammation plays a limited role in the regulation of adipose tissue COL1A1 protein abundance. *J. Nutr. Biochem.* **2012**, *23*, 567–572. [CrossRef] [PubMed]

80. San, J.; Du, Y.; Wu, G.; Xu, R.; Yang, J.; Hu, J. Transcriptome analysis identifies signaling pathways related to meat quality in broiler chickens–the extracellular matrix (ECM) receptor interaction signaling pathway. *Poult. Sci.* **2021**, *100*, 101135. [CrossRef] [PubMed]
81. Lee, H.-J.; Jang, M.; Kim, H.; Kwak, W.; Park, W.; Hwang, J.Y.; Lee, C.-K.; Jang, G.W.; Park, M.N.; Kim, H.-C. Comparative transcriptome analysis of adipose tissues reveals that ECM-receptor interaction is involved in the depot-specific adipogenesis in cattle. *PLoS ONE* **2013**, *8*, e66267. [CrossRef]
82. Rosso, F.; Giordano, A.; Barbarisi, M.; Barbarisi, A. From cell–ECM interactions to tissue engineering. *J. Cell. Physiol.* **2004**, *199*, 174–180. [CrossRef] [PubMed]
83. Franchini, M.; Monnais, E.; Seboek, D.; Radimerski, T.; Zini, E.; Kaufmann, K.; Lutz, T.; Reusch, C.; Ackermann, M.; Muller, B. Insulin resistance and increased lipolysis in bone marrow derived adipocytes stimulated with agonists of Toll-like receptors. *Horm. Metab. Res.* **2010**, *42*, 703–709. [CrossRef]
84. Suganami, T.; Tanimoto-Koyama, K.; Nishida, J.; Itoh, M.; Yuan, X.; Mizuarai, S.; Kotani, H.; Yamaoka, S.; Miyake, K.; Aoe, S. Role of the Toll-like receptor 4/NF-κB pathway in saturated fatty acid–induced inflammatory changes in the interaction between adipocytes and macrophages. *Arterioscler. Thromb. Vasc. Biol.* **2007**, *27*, 84–91. [CrossRef]
85. Zu, L.; He, J.; Jiang, H.; Xu, C.; Pu, S.; Xu, G. Bacterial endotoxin stimulates adipose lipolysis via toll-like receptor 4 and extracellular signal-regulated kinase pathway. *J. Biol. Chem.* **2009**, *284*, 5915–5926. [CrossRef] [PubMed]
86. Purohit, J.S.; Hu, P.; Chen, G.; Whelan, J.; Moustaid-Moussa, N.; Zhao, L. Activation of nucleotide oligomerization domain containing protein 1 induces lipolysis through NF-κB and the lipolytic PKA activation in 3T3-L1 adipocytes. *Biochem. Cell Biol.* **2013**, *91*, 428–434. [CrossRef] [PubMed]
87. Ye, S.; Ding, Y.-F.; Jia, W.-H.; Liu, X.-L.; Feng, J.-Y.; Zhu, Q.; Cai, S.-L.; Yang, Y.-S.; Lu, Q.-Y.; Huang, X.-T. SET domain–containing protein 4 epigenetically controls breast cancer stem cell quiescence. *Cancer Res.* **2019**, *79*, 4729–4743. [CrossRef] [PubMed]
88. Graja, A.; Garcia-Carrizo, F.; Jank, A.M.; Gohlke, S.; Ambrosi, T.H.; Jonas, W.; Ussar, S.; Kern, M.; Schürmann, A.; Aleksandrova, K. Loss of periostin occurs in aging adipose tissue of mice and its genetic ablation impairs adipose tissue lipid metabolism. *Aging Cell* **2018**, *17*, e12810. [CrossRef]

animals MDPI

Article

Genotyping-by-Sequencing Strategy for Integrating Genomic Structure, Diversity and Performance of Various Japanese Quail (*Coturnix japonica*) Breeds

Natalia A. Volkova [1,†], Michael N. Romanov [1,2,*,†], Alexandra S. Abdelmanova [1], Polina V. Larionova [1], Nadezhda Yu. German [1], Anastasia N. Vetokh [1], Alexey V. Shakhin [1], Ludmila A. Volkova [1], Dmitry V. Anshakov [3], Vladimir I. Fisinin [4], Valeriy G. Narushin [5,6], Darren K. Griffin [2], Johann Sölkner [7], Gottfried Brem [8], John C. McEwan [9], Rudiger Brauning [9] and Natalia A. Zinovieva [1,*]

1 L. K. Ernst Federal Research Center for Animal Husbandry, Dubrovitsy, Podolsk 142132, Moscow Oblast, Russia; natavolkova@inbox.ru (N.A.V.); abdelmanova@vij.ru (A.S.A.); volpolina@mail.ru (P.V.L.); ngerman9@gmail.com (N.Y.G.); anastezuya@mail.ru (A.N.V.); alexshahin@mail.ru (A.V.S.); ludavolkova@inbox.ru (L.A.V.)
2 School of Biosciences, University of Kent, Canterbury, Kent CT2 7NJ, UK; d.k.griffin@kent.ac.uk
3 Breeding and Genetic Center Zagorsk Experimental Breeding Farm—Branch of the Federal Research Centre, All-Russian Poultry Research and Technological Institute, Russian Academy of Sciences, Sergiev Posad 141311, Moscow Oblast, Russia; a89265594669@rambler.ru
4 Federal Research Center "All-Russian Poultry Research and Technological Institute" of the Russian Academy of Sciences, Sergiev Posad 141311, Moscow Oblast, Russia; olga@vnitip.ru
5 Research Institute for Environment Treatment, 69032 Zaporizhya, Ukraine; val@vitamarket.com.ua
6 Vita-Market Co., Ltd., 69032 Zaporizhya, Ukraine
7 Institute of Livestock Sciences (NUWI), University of Natural Resources and Life Sciences Vienna, 1180 Vienna, Austria; johann.soelkner@boku.ac.at
8 Institute of Animal Breeding and Genetics, University of Veterinary Medicine, 1210 Vienna, Austria; gottfried.brem@agrobiogen.de
9 AgResearch, Invermay Agricultural Centre, Mosgiel 9053, New Zealand; john.mcewan@agresearch.co.nz (J.C.M.); rudiger.brauning@agresearch.co.nz (R.B.)
* Correspondence: m.romanov@kent.ac.uk (M.N.R.); n_zinovieva@mail.ru (N.A.Z.)
† These authors contributed equally to this work.

Simple Summary: Artificial selection has been applied to domesticated birds for many decades. More recently, this selection has made use of so-called single-nucleotide polymorphism (SNP) markers—simple variants in a DNA sequence. These SNPs can be used for whole-genome screening to detect the unique traces of areas of the genome that are subject to selection. Doing this may help to shed light on the evolutionary and family history (phylogeny) of domestic Japanese quails of different breeds and utility types (e.g., egg, meat or dual-purpose breeds). In this study, 99 birds were used, representing eight breeds (11% of the world's quail gene pool) and various purposes of use to gather genetic (whole-genome) data in the first-ever analysis of its kind performed on domestic quails. We thereby uncovered evolutionary relationships and points of divergence of individual quail breeds, gleaning important insights into the genetic diversity of domestic quail breeds and their future breeding potential.

Abstract: Traces of long-term artificial selection can be detected in genomes of domesticated birds via whole-genome screening using single-nucleotide polymorphism (SNP) markers. This study thus examined putative genomic regions under selection that are relevant to the development history, divergence and phylogeny among Japanese quails of various breeds and utility types. We sampled 99 birds from eight breeds (11% of the global gene pool) of egg (Japanese, English White, English Black, Tuxedo and Manchurian Golden), meat (Texas White and Pharaoh) and dual-purpose (Estonian) types. The genotyping-by-sequencing analysis was performed for the first time in domestic quails, providing 62,935 SNPs. Using principal component analysis, Neighbor-Net and Admixture algorithms, the studied breeds were characterized according to their genomic architecture, ancestry and direction of selective breeding. Japanese and Pharaoh breeds had the smallest number and length of homozygous segments indicating a lower selective pressure. Tuxedo and Texas White breeds showed the highest values of these indicators and genomic inbreeding suggesting a greater homozygosity. We revealed evidence for the integration of genomic and performance data, and our findings are applicable for elucidating the history of creation and genomic variability in quail breeds that, in turn, will be useful for future breeding improvement strategies.

Keywords: genotyping-by-sequencing; genetic diversity; genomic structure; phylogeny; performance; Japanese quail; breeds; utility types

1. Introduction

The study of molecular genetic principles that determine the degree of manifestation of economically significant traits is of crucial importance for increasing agricultural. In so doing, it can be applied to produce the most effective and cost-efficient agricultural products for domestic and world consumption. The poultry industry is one of the key sectors of agricultural production, and its important products are meat and eggs. Poultry meat accounts for 45% of the total global meat production [1] and virtually all egg production. In Russia, the growing market demand for poultry products led, by 2020, to an elevation from 4 to 10% in the share of food products from non-traditional poultry species [2]. An increasing proportion of the world's egg and meat supply are provided from species of the Phasianidae (pheasant) family [3,4]. Hereby, quail breeding industry products are in special demand worldwide because of the palatability of quail eggs and meat, as well as the early onset of sexual maturity. Because of this, the establishment and growth of large quail farms mean that quail eggs and meat are becoming everyday products [3–7].

The progenitor of contemporary quail breeds, the Japanese quail (*Coturnix japonica* Temminck & Schlegel, 1848), is a migratory bird native to East Asia. Domesticated Japanese quail are a common poultry type used for meat and eggs in Europe, Asia and throughout the world. Quails have been used in genetic research since 1940 [8] and, over time, they have become an increasingly important biological model for developmental, behavioral and biomedical studies [3,4]. Belonging to the same Phasianidae family as chickens, quails have a number of advantages as a research model. They are small, fast growing, and have a short life cycle, reaching sexual maturity in 7–8 weeks after hatching [9]. In comparative biological studies of galliforms, quails show key differences from chickens and some other poultry species, e.g., immune status, migratory and seasonal behavior [3].

Worldwide, there are about 70 domestic Japanese quail breeds or strains, including commercial and laboratory quails [10]. The first Japanese quails were imported into Russia in 1964 for breeding and production purposes. The adult quail population in the former USSR grew steadily year after year, peaking at around 200,000 individual birds [11]. A large collection of quail breeds was first created at the Moscow Timiryazev Agricultural Academy (MTAA) and involved the Pharaoh (PHA), English White (ENW), British Range, Tuxedo (TUX) and Marbled breeds [12] plus a wild-type colored strain developed in the Scientific and Production Association "Complex" (SPAC; Moscow, Russia) by crossing the Marbled and PHA quails. The Marbled breed was created at the MTAA in collaboration with the N.I. Vavilov Institute of General Genetics by subjecting a group of quails to X-rays. A relatively novel dual-purpose Estonian (EST) breed was produced in 1988 by mating the Japanese (JAP), ENW and PHA breeds. According to a cytogenetic analysis, the EST quails can be distinguished from JAP quails by the presence of a centromeric band in autosome 1 that is G-positive and can be utilized as a chromosomal marker for EST (as reviewed in [11]). Another large collection of quail breeds currently exists in the Zagorsk Experimental Breeding Farm, All-Russian Poultry Research and Technological Institute (ZEBF/ARPRTI), and embraces JAP, ENW, English Black (ENB), TUX, Manchurian Golden (MAG), EST, PHA, Texas White (TEW) and a few other quail breeds and strains [13].

To create a competitive breeder stock for quails, it is necessary to use modern methodologies of genetic and genomic analysis aimed at increasing the efficiency of selection and breeding work. For instance, identified sex-linked genes for phenotypic traits, e.g., recessive genes for imperfect albinism (*al*) and brown (*br*), can be employed for autosexing (sex sorting) of newly hatched chicks [10,11,14,15].

The integration of high-throughput, next-generation sequencing-based genomic technologies into practical quail breeding should therefore be carried out, at the initial stage, by assessing the genomic architecture characteristic of a particular breed. A subsequent comparison of the genomic structure of quail breeds of different origin and direction of selective breeding is necessary. This can prove to be an effective approach to identify specific genomic regions that are either related to recent divergence and/or earlier breeding differentiation. In addition, such an investigation facilitates a genome-wide assessment and refinement of the diversity and phylogeny amongst various quail breeds. Genotyping by sequencing (GBS), one of the restriction enzyme-based enrichment approaches designed initially for plants [16], is a promising strategy for reducing the financial burden of selection strategies via high sample multiplexing, focusing the sequenced genome areas on randomly distributed read tags [17]. Being a relatively affordable and widely applicable substitute for concurrent single-nucleotide polymorphism (SNP) mining and genotyping in plants (e.g., [16,18,19]), GBS is also increasingly being used in animals (e.g., [17,19–22]). Indeed, there was a recent report that GBS has been utilized for evaluating demographic history and genetic divergence in wild African harlequin quail (*Coturnix delegorguei delegorguei*) populations of Kenya [22]. Ravagni et al. [23] employed GBS to explore the evolutionary history of an island endemic, the common quail (*Coturnix coturnix*) in the Azores archipelago. To the best of our knowledge, however, there have been no studies using this technique to characterize the genomic architecture and diversity in the divergently selected breeds of the domesticated Japanese quail.

In the current investigation, we thus aimed to examine and compare the variability and phylogeny among the genomes of eight divergently selected breeds of egg-type, meat-type and dual-purpose quails that represented a significant portion (~11%) of the global gene pool of quail breeds. In accordance with this goal, the GBS approach was applied to characterize the genetic structure of these quail populations. This information is essential for maintaining their genomic diversity and facilitating efficient breeding in the future.

2. Materials and Methods

2.1. Experimental Birds and Performance Data

Quails were hatched from fertile eggs purchased from the Genofond LLC (ZEBF/ARPRTI; [13]), grown at the L. K. Ernst Federal Research Centre for Animal Husbandry (LKEFRCAH) [24], and sampled for DNA. The following eight quail breeds were used in this experiment (Table 1): JAP, ENW, ENB, TUX, and MAG (of egg type); TEW and PHA (of meat type); and EST (of dual purpose).

Table 1. Eight quail breeds involved in this study.

Breed	Code	n [1]	Origin	Refs
		Egg type		
Japanese 	JAP	19	Japan; domesticated in Japan and China in 12th century or earlier; selected in the 1st half of the 20th century, brought to the USSR from Japan in the mid-20th century and/or from Yugoslavia in 1964	[10–13,25–27]

Table 1. *Cont.*

Breed	Code	n [1]	Origin	Refs
English (British) White 	ENW	11	England; a mutant from JAP quails; imported to the USSR from Hungary in 1987	[12,13,24,27]
English (British) Black 	ENB	13	England; a mutant from JAP quails; imported to the USSR from Hungary in 1971	[13,27]
Tuxedo 	TUX	16	from crossing ENW and ENB	[12,13,27]
Manchurian (Manchu) Golden (or Golden Phoenix) 	MAG	14	Marsh Farms, CA, USA, 1960s; bred by Albert Marsh as a natural mutant in a flock of brown-colored quails	[12,13,27–29]
		Dual purpose (or universal)		
Estonian (or Kitevers) 	EST	9	Estonia, 1988; from crossing JAP (a Moscow line), ENW and Pharaoh	[11,13,27]
		Meat type		
Pharaoh 	PHA	10	USA; wild-type plumage; an imported French fattening line used in this study	[12,13,26,27,30]
Texas White (or Texas Pharaoh, White Pharaoh, Snowy) 	TEW	7	Texas, USA; from crossing PHA and ENW	[27,30]

[1] n, number of individuals after quality control.

For each breed, the number of females (*n*) was taken into account, for which the following performance indicators (as mean ± standard deviation) were collected: egg number (EN) for 180 days from the start of lay; egg weight (EW) obtained at the age of 180 to 210 days (for each female, mean was calculated over all eggs laid during a given period); and body weight (BW) of females at the ages of 6 weeks and 6 months. These data were subsequently assessed and compared with calculations of interbreed genetic variability resulting from the GBS analysis. Herewith, we proposed a hypothesis that a certain degree of "congruence" (or, in other words, integration) between phenotypic and genomic data can take place for this sample of quail breeds. For this purpose, an appropriate mathematical analysis was undertaken using a new index, Narushin's IPI (Integral Performance Index). The latter was recently established by Vakhrameev et al. [31] to evaluate the main economically important traits (i.e., EN, EW and female BW) in various chicken breeds and was originally designated as EY/W (where EY was the product of mean EN and EW, and W was mean female BW). Here, we renamed this index after the author of that study, Valeriy G. Narushin, who proposed this indicator, and calculated it using the corresponding formula:

$$\mathrm{IPI} = \frac{\mathrm{EN \cdot EW}}{\mathrm{BW}}$$

where EN is egg number, EW is egg weight (in g), and BW is female body weight at 6 months of age (in g), all values being calculated as breed means.

Statistical evaluation of raw performance data and means was performed using Microsoft Excel (version 16.66.1). Student's *t*-test was implemented to compare the means of breeds in pairwise mode and determine the significance of the differences between them using Microsoft Excel's T.TEST function and GraphPad online calculator [32].

2.2. Sampling and DNA Isolation

Feather samples containing pulp were obtained from 106 quails of all the breeds studied. DNA extraction was performed using the Syntol kit for DNA isolation from animal tissues (Syntol, Moscow, Russia). The DNA solution concentration was determined using a Qubit 3.0 fluorimeter (Thermo Fisher Scientific, Wilmington, DE, USA). To check the purity of the extracted DNA, the OD260/280 ratio was tested using a NanoDrop-2000 instrument (Thermo Fisher Scientific).

2.3. Sequencing, Genotyping and Quality Control of SNPs

Quail genotyping was performed using GBS analysis [16] that included the basic steps of library construction, sequencing, sequence quality control (QC), SNP detection, and construction of a genomic relationship matrix. In particular, the methods described in Elshire et al. [16], with changes as in Dodds et al. [33], were implemented to build the GBS libraries. A *Pst*I–*Msp*I double-digest was used to generate one GBS library that also contained negative control samples devoid of DNA. Libraries were subjected to a Pippin Prep (SAGE Science, Beverly, MA, USA) to choose fragments with a size between 220 and 340 bp (genomic sequence plus 148 bp adapters). We employed a set of 768 barcodes designed by Integrated DNA Technologies, Inc. (Coralville, IA, USA) and Illumina (Illumina, Inc., San Diego, CA, USA) that differed from each other by at least three mutational steps. The corresponding adapter sequences were as follows: *Pst*I_Common_F, AGATCGGAAGAGCACACGTCTGAACTCCAGTCAC; *Pst*I_Common_R, GTGACTGGAGTTCAGACGTGTGCTCTTCCGATCTTGCA; *Msp*I(Y)_Common_F, CGAGATCGGAAGAGCGGACTTTAAGC; and *Msp*I_Common_R, GTGACTGGAGTTCAGACGTGTGCTCTTCCGATCT. Single-end sequencing (1 × 101 bp) was performed utilizing a NovaSeq 6000 instrument (Illumina, Inc.) and the appropriate v1.5 reagents. Raw fastq files were quality checked using a custom QC pipeline, DECONVQC [19,34]. As one of the QC steps, raw fastq files were quality tested using FastQC [35].

As a reference genome, we used the Japanese quail genome assembly Coturnix japonica 2.0 [36], along with the databases Ensembl 104 and Ensembl Genomes 51 (released on 7 May 2021; [37]). Removal of adapter sequences and demultiplexing of the fastq file, i.e., its separation by samples to produce individual fastq files using a list of barcodes, were executed using the cutadapt program [38,39]. The QC of fastq files was carried out in the FastQC program [35]. To call SNPs from the GBS data, the bioinformatics workflow snpGBS [40,41] was employed. The bowtie2 package was used to align the individual fastq files to the reference genome and index them [42], while sorting of bam files was performed using samtools [43,44].

Joint genotyping of the resulting files was implemented using bcftools [44] generating one multi sample VCF file. After filtering, 80,673 SNPs were used for subsequent analysis steps. The data was generated into a file format acceptable for further analysis using the R software package [45]. The PLINK 1.9 program [46] was employed to control the quality of SNP detection. The obtained quail genotypes were filtered according to the genotyping efficiency parameter (mind 0.25), and SNPs genotyped in less than 90% samples (geno 0.1) were excluded from the analysis. The final dataset used for the genome-wide analyses was 62,935 SNPs out of original 80,673 SNPs. A total of 106 individuals were initially sequenced and genotyped. After removing two quails that did not belong to their respective breeds according to genetic data, a matrix of 104 birds was subject to subsequent analysis. After pulling out five more quails from the total number due to insufficient information about them with regard to the studied SNPs, a total of 99 individuals was investigated in the experiment.

For some types of analysis, e.g., principal component analysis (PCA), analysis of genetic diversity and divergence, construction of phylogenetic networks, analysis of population structure and analysis of gene flow (migration events), an additional linkage disequilibrium (LD) filter was applied to remove loci for which LD was identified within a 50 Kb sliding window with a step of 5. After using the LD filter, 27,171 SNPs were included in the analysis.

2.4. Genetic Diversity Assessment

To determine within-population genetic diversity, PLINK 1.9 software package was used. QC was performed at both the individual and SNP levels using PLINK 1.9. By implementing various parameters of the R package diveRsity [47], we computed values of observed heterozygosity (H_O), expected heterozygosity (H_E), unbiased expected heterozygosity ($_UH_E$), rarefied allelic richness (A_R; [48]), coefficient inbreeding (F_{IS}) and coefficient of inbreeding ($_UF_{IS}$) based on unbiased expected heterozygosity.

2.5. PCA, Neighbor-Net and Admixture Procedures

For breed clustering, PCA and calculation of identical-by-state (IBS) distances were performed in PLINK1.9. The degree of genetic differentiation of the studied breeds was estimated based on pairwise F_{ST} values. Visualization of PCA results was carried out in the R ggplot2 package [49]. Dendrograms based on IBS distances and pairwise F_{ST} distances were plotted using an agglomerative method for constructing phylogenetic networks, i.e., Neighbor-Net in SplitsTree 4 [50]. The software Admixture v1.3 [51] for model-based clustering and computation of the related cross-validation (CV) errors was implemented to analyze ancestral populations and genetic impurities, while the BITE R package [52] was used to visualize these results. The Phantasus web program was also used to perform PCA and hierarchical clustering procedures [53]. Using the online T-REX program [54], the Neighbor-Joining [55] trees showing phylogenetic relationships between breeds were built.

Gene flow (migration) events were analyzed using the TreeMix 1.12 program [56]. The analysis considered from 0 to 5 migrations with 30 iterations per migration event. The optimal number of migrations (1) was determined using the *OptM* R package [57]. The best maximum likelihood tree configuration was determined based on the minimum mean standard error of the residual matrix among all iterations.

The analysis of homozygous genomic segments (runs of homozygosity, ROH) was performed using the detectRUNS R package [58] with the following settings: the minimum number of SNPs was 30, and the minimum length was 0.5 Mb.

3. Results

3.1. Breed Performance

Information for the three major performance characteristics, according to which the IPI index was computed, is given in Table 2. As can be seen, the two meat-type breeds, PHA and TEW, had the lowest IPI values (roughly 5 if rounded up to integers), the dual-purpose breed, EST, had a slightly higher value (~7), and the five egg-type breeds had greater values (~8 to 12).

A matrix of interbreed Euclidean distances computed for breed IPI values is presented in Supplementary Table S1. Using it, breed clustering was reconstructed in the form of PCA plots and Neighbor-Joining trees (Figure 1). Notably, a largely similar configuration of breeds was obtained using both clustering techniques, reflecting the breed subdivision into three main types of utility and selection, as well as in accordance with the ranking of IPI values as shown in Table 2.

3.2. Analysis of Genetic Diversity

Using around 100 DNA samples from quails of as many different phenotypes as possible, we generated a GBS panel for genotyping the quail breeds. In terms of genetic diversity values (Table 3), TUX quails were characterized by lower values of genetic diversity, as measured by lower levels of expected heterozygosity (H_E = 0.263 vs. the maximum value of 0.310, $p < 0.001$) and allelic richness (A_R = 1.730 vs. the maximum value of 1.864, $p < 0.001$) as compared to the JAP breed. This can be explained by the higher intensity of breeding work in TUX that was aimed at consolidating the desired breed characteristics. The inbreeding coefficient (F_{IS}) was represented by the maximum value for the JAP population (0.020, with a 95% confidence interval being from 0.016 to 0.024), which may be indicative of a likely growth in gene homozygosity in this population.

Due to the small number of birds in each breed, we also calculated unbiased measures of expected heterozygosity ($_{U}H_E$) and expected inbreeding rate ($_{U}F_{IS}$) adjusted for small samples. The former was highest in EST (0.313) and JAP (0.319). The latter coefficient was represented by positive values for all breeds ranging from the minimum of 0.011 in TEW to the maximum of 0.046 in JAP quails. High rates of $_{U}F_{IS}$ were also found in EST (0.032) and MAG (0.031), as well as in the PHA population (0.035). This enabled us to conclude that there was a significantly higher homozygosity of genes in these four populations as compared to other breeds.

3.3. Between-Breed Genetic Relationships and Model-Based Clustering

PCA plots for various eight quail breeds based on the individual nucleotide sequence data as obtained using the GBS method are graphically presented in Figure 2.

The first component was responsible for 44.42% of genetic variability and differentiated the cluster of ENW, ENB and TUX from the cluster of JAP, EST, PHA and TEW. The second component conformed to 23.48% of genetic differences and showed the remote position of MAG relative to the other quail breeds, i.e., demonstrated its isolation from them. In the PC1–PC3 plane, TEW differentiated from the rest, and in the PC1–PC2 plane, the MAG population was separated from the others.

A Neighbor-Net tree based on the matrix of pairwise IBS distances for different quail population individuals revealed a clearcut breed differentiation judging from the distribution of individuals relative to each other in Figure 3.

Table 2. Performance indicators [1] of females from the eight quail breeds studied (mean ± SD).

Breed [2]	n	EN	EW	BW		IPI
				6 Weeks	6 Months	
			Egg type			
JAP (a)	41	165.4 ± 14.7 ^{a}	11.0 ± 0.9 ^{a}	146.7 ± 13.6 ^{a}	149.0 ± 13.2 ^{a}	12.2 ± 1.2 ^{a}
ENW (b)	11	134.8 ± 8.5 ^{a,b}	10.2 ± 1.0 ^{a,b}	157.6 ± 12.7 ^{a,b}	166.6 ± 9.0 ^{a,b}	8.3 ± 1.2 ^{a,b}
ENB (c)	11	133.4 ± 8.0 ^{a,c}	10.4 ± 0.9 ^{c}	151.5 + 15.0 ^{c}	159.5 ± 14.0 ^{a,c}	8.8 ± 1.4 ^{a,c}
TUX (d)	11	131.1 ± 7.4 ^{a,d}	10.2 ± 0.9 ^{a,d}	141.4 ± 10.5 ^{b,d}	149.0 ± 12.6 ^{b,d}	9.0 ± 0.8 ^{a,d}
MAG (e)	12	147.5 ± 4.5 ^{a,b,c,d,e}	10.6 ± 1.6 ^{e}	168.2 ± 17.2 ^{a,c,d,e}	180.1 ± 18.9 ^{a,b,c,d,e}	8.8 ± 2.1 ^{a,e}
			Dual purpose			
EST (f)	18	148.9 ± 9.7 ^{a,b,c,d,f}	11.8 ± 1.4 ^{a,b,c,d,e,f}	248.2 ± 10.6 ^{a,b,c,d,e,f}	247.3 ± 15.4 ^{a,b,c,d,e,f}	7.1 ± 1.1 ^{a,b,c,d,e,f}
			Meat type			
PHA (g)	12	118.2 ± 8.2 ^{a,b,c,d,e,f}	12.6 ± 0.8 ^{a,b,c,d,e,f}	292.3 ± 16.2 ^{a,b,c,d,e,f,g}	294.3 ± 19.5 ^{a,b,c,d,e,f,g}	5.1 ± 0.6 ^{a,b,c,d,e,f}
TEW (h)	23	121.4 ± 18.7 ^{a,b,c,d,e,f}	12.7 ± 1.0 ^{a,b,c,d,e,f}	305.5 + 21.3 ^{a,b,c,d,e,f,g}	317.7 ± 25.9 ^{a,b,c,d,e,f,g}	4.9 ± 1.0 ^{a,b,c,d,e,f}

[1] n, number of individuals; EN, egg number; EW, egg weight (g); BW, female body weight at 6 weeks and 6 months of age (g); IPI, Integral Performance Index. [2] Quail breeds: JAP, Japanese; ENW, English White; ENB, English Black; TUX, Tuxedo; MAG, Manchurian Golden; EST, Estonian; PHA, Pharaoh; TEW, Texas White. (a–h) Significant pairwise differences for breeds with the corresponding same superscript ($p < 0.05$); the absence of a corresponding common superscript indicates that the differences between the specific two breeds are insignificant.

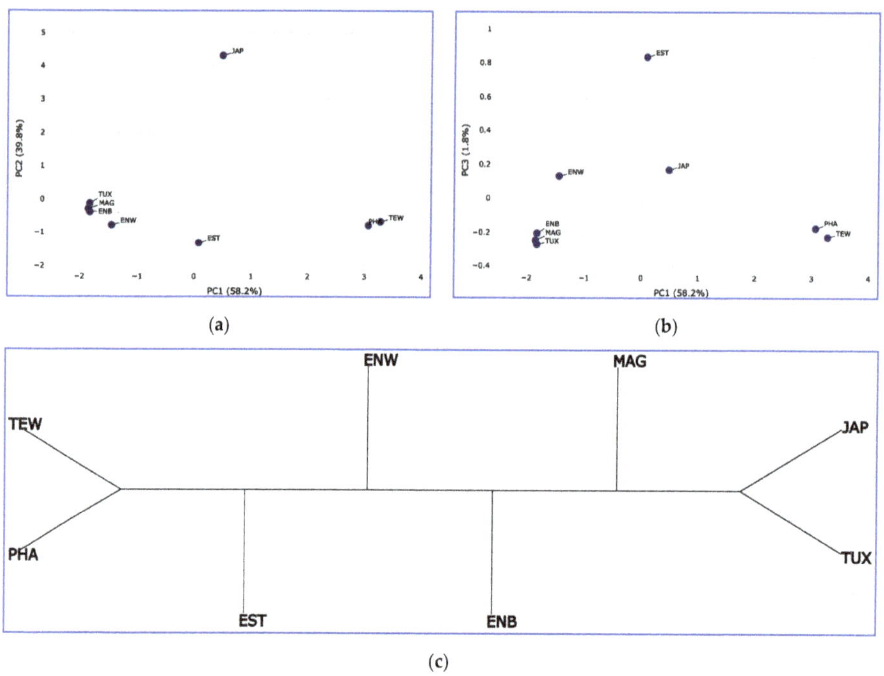

Figure 1. Clustering reconstruction of the eight breeds studied using the IPI-based pairwise Euclidean distances. (**a**,**b**) PCA plots for first (PC1) and second (PC2) components (**a**), and for first (PC1) and third (PC3) components (**b**) using the Phantasus web tool [53]. (**c**) A Neighbor-Joining rootless axial tree built with no proportional edge length and using the Neighbor Joining method [55] and the online T-REX tool [54]. Quail breeds: JAP, Japanese; ENW, English White; ENB, English Black; TUX, Tuxedo; MAG, Manchurian Golden; EST, Estonian; PHA, Pharaoh; TEW, Texas White.

Table 3. Characterization of the genetic diversity parameters [1] in the quail populations studied.

Breed [2]	H_O (M ± SE)	H_E (M ± SE)	$_uH_E$ (M ± SE)	A_R (M ± SE)	F_{IS} [CI 95%]	$_uF_{IS}$ [CI 95%]
			Egg type			
JAP	0.303 ± 0.001	0.310 ± 0.001	0.319 ± 0.001	1.864 ± 0.001	0.020 [0.016; 0.024]	0.046 [0.043; 0.049]
ENW	0.281 ± 0.001	0.273 ± 0.001	0.287 ± 0.001	1.778 ± 0.002	−0.029 [−0.033; −0.025]	0.020 [0.016; 0.024]
ENB	0.282 ± 0.001	0.276 ± 0.001	0.287 ± 0.001	1.774 ± 0.002	−0.019 [−0.023; −0.015]	0.020 [0.016; 0.024]
TUX	0.265 ± 0.001	0.263 ± 0.001	0.271 ± 0.001	1.730 ± 0.002	−0.010 [−0.014; −0.006]	0.022 [0.018; 0.026]
MAG	0.286 ± 0.001	0.285 ± 0.001	0.295 ± 0.001	1.790 ± 0.002	−0.005 [−0.009; −0.001]	0.031 [0.027; 0.035]
			Dual purpose			
EST	0.302 ± 0.001	0.295 ± 0.001	0.313 ± 0.001	1.839 ± 0.002	−0.025 [−0.030; −0.020]	0.032 [0.028; 0.036]
			Meat type			
PHA	0.290 ± 0.001	0.286 ± 0.001	0.301 ± 0.001	1.815 ± 0.002	−0.017 [−0.021; −0.013]	0.035 [0.031; 0.039]
TEW	0.282 ± 0.002	0.264 ± 0.001	0.284 ± 0.001	1.757 ± 0.003	−0.067 [−0.072; −0.062]	0.011 [0.006; 0.016]

[1] H_O, observed heterozygosity; M, mean value; SE, standard error; H_E, expected heterozygosity; $_uH_E$, unbiased expected heterozygosity adjusted for small samples; A_R, rarefied allelic richness; F_{IS}, inbreeding coefficient [CI 95%, range variation of F_{IS} coefficient at a confidence interval of 95%]; $_uF_{IS}$, unbiased inbreeding coefficient [CI 95%, range variation of $_uF_{IS}$ coefficient at a confidence interval of 95%] adjusted for small samples. [2] Quail breeds: JAP, Japanese; ENW, English White; ENB, English Black; TUX, Tuxedo; MAG, Manchurian Golden; EST, Estonian; PHA, Pharaoh; TEW, Texas White.

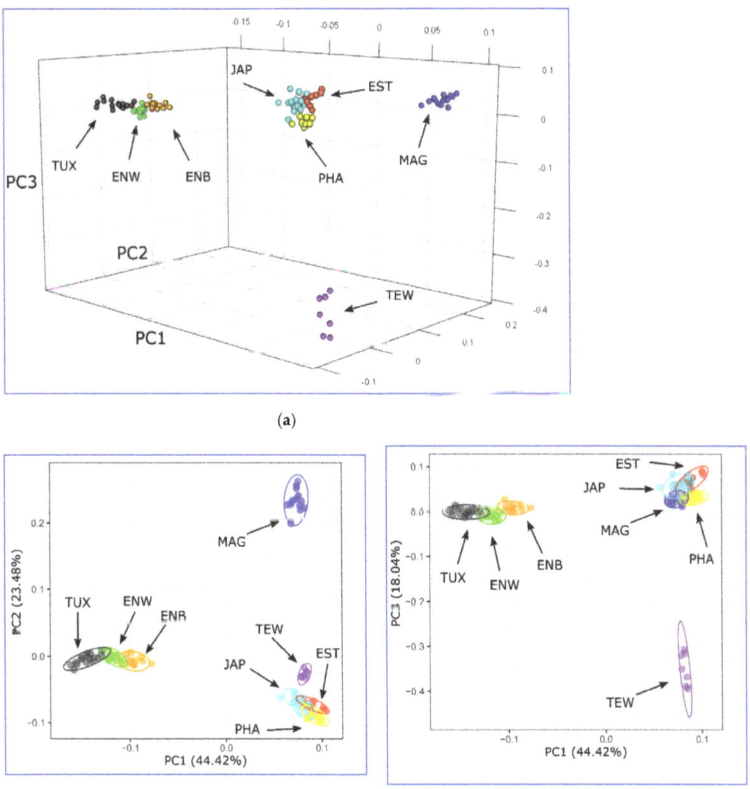

(a)

(b)

(c)

Figure 2. GBS-based PCA plots for the eight quail breeds studied. (**a**) View in 3D. (**b**) Plot composed in the plane of the first (*X*-axis, PC1) and second (*Y*-axis, PC2) components. (**c**) Plot drawn in the plane of the first (*X*-axis, PC1) and third (*Y*-axis, PC3) components. Quail breeds: JAP, Japanese; ENW, English White; ENB, English Black; TUX, Tuxedo; MAG, Manchurian Golden; EST, Estonian; PHA, Pharaoh; TEW, Texas White.

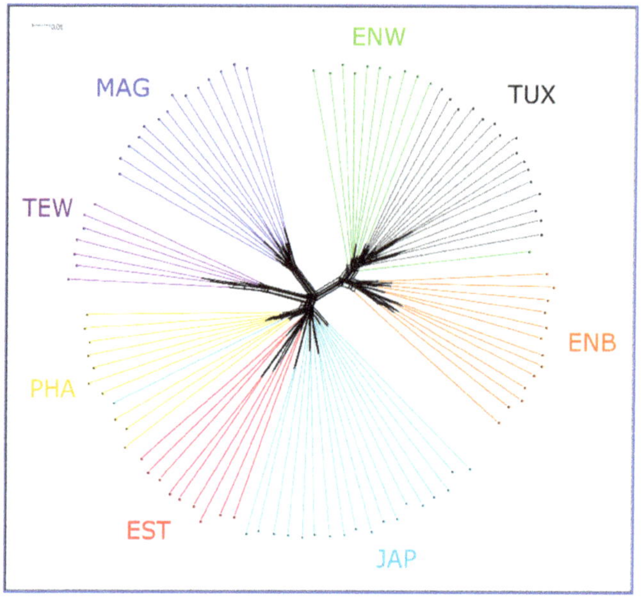

Figure 3. Neighbor-Net tree based on pairwise IBS distances. Quail breeds: JAP, Japanese; ENW, English White; ENB, English Black; TUX, Tuxedo; MAG, Manchurian Golden; EST, Estonian; PHA, Pharaoh; TEW, Texas White.

During the Admixture-assisted analysis of ancestor populations and genetic impurities, calculations of the CV error for a different number of clusters (from 1 to 9) showed that the optimal number of clusters (K) was equal to 3 (Figure 4a,b). At K = 5, two groups of populations were clearly distinguished from each other as follows: (1) ENB + ENW + TUX, and (2) JAP + EST + PHA, while two single breeds, MAG and TEW, were genetically unique and distinct (Figure 4b). Clustering in the Admixture program (Figure 4c) demonstrated that MAG at K = 3 and TEW at K = 4 formed their own specific genomic pattern that was not observed in the other breeds. At the maximum tested level of clustering (K = 9), it was found that the genomic components predominantly represented in PHA were also present in EST and JAP, although this was already pronounced to a much lesser extent when K equaled 6 to 9. Shared genetic components were also observed in ENB (K = 4), ENW (K = 6 and K = 8), and TUX (K = 9).

The ENB, ENW and TUX breeds that formed a separate cluster partially overlapped each other. This was consistent with the history of the TUX descent through crossbreeding between ENW and English ENB, as well as the selection of these breeds for egg production. The formation of a joint cluster of JAP, EST, TEW and PHA breeds also conformed to the history of their origin and breeding. In particular, when creating EST, the PHA, JAP and ENW breeds were involved, and when developing TEW, the PHA breed was used.

To visualize the genetic distances between the studied populations, a dendrogram of phylogenetic networks were constructed based on pairwise F_{ST} genetic distances (Supplementary Table S2) and using the Neighbor-Net algorithm (Figure 5a).

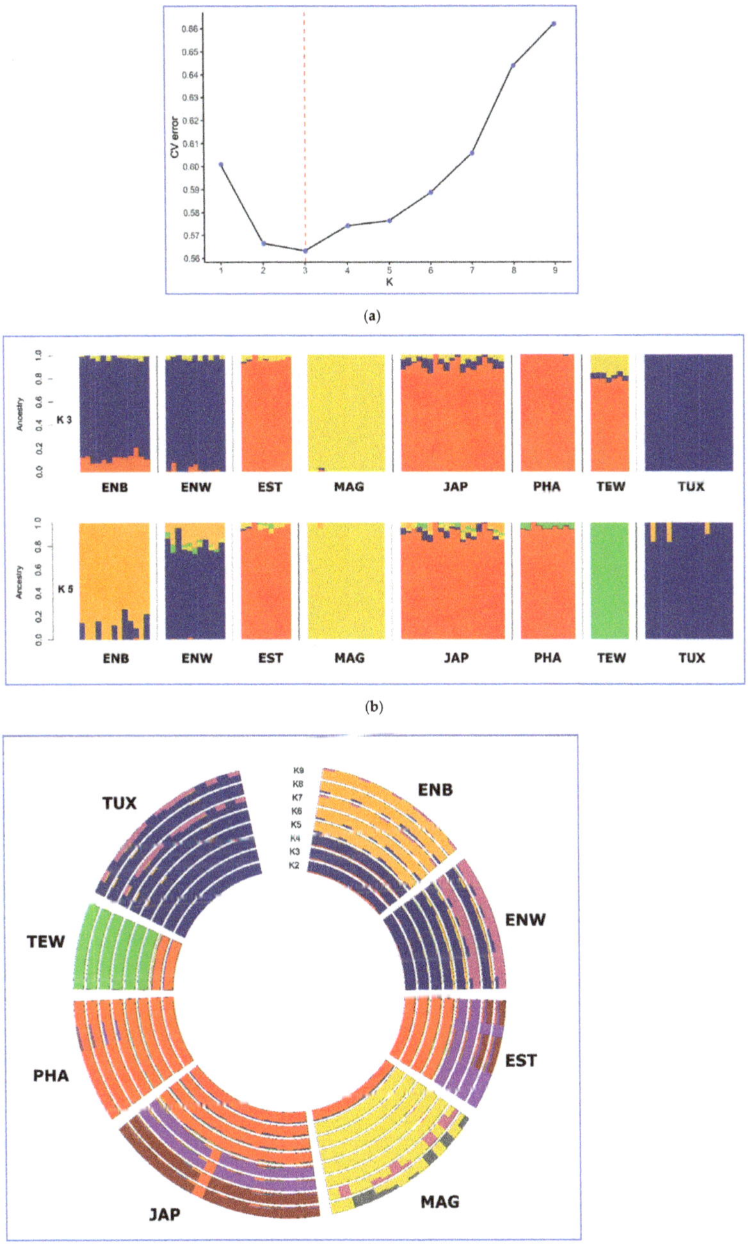

Figure 4. Admixture-assisted ancestry cluster analysis. (**a**) CV error calculations for different number of ancestral populations or clusters (from 1 to 9). (**b**) Horizontal view at K = 3 and 5 clusters. (**c**) Circular view for K equaling 2 to 9 clusters. Quail breeds: JAP, Japanese; ENW, English White; ENB, English Black; TUX, Tuxedo; MAG, Manchurian Golden; EST, Estonian; PHA, Pharaoh; TEW, Texas White.

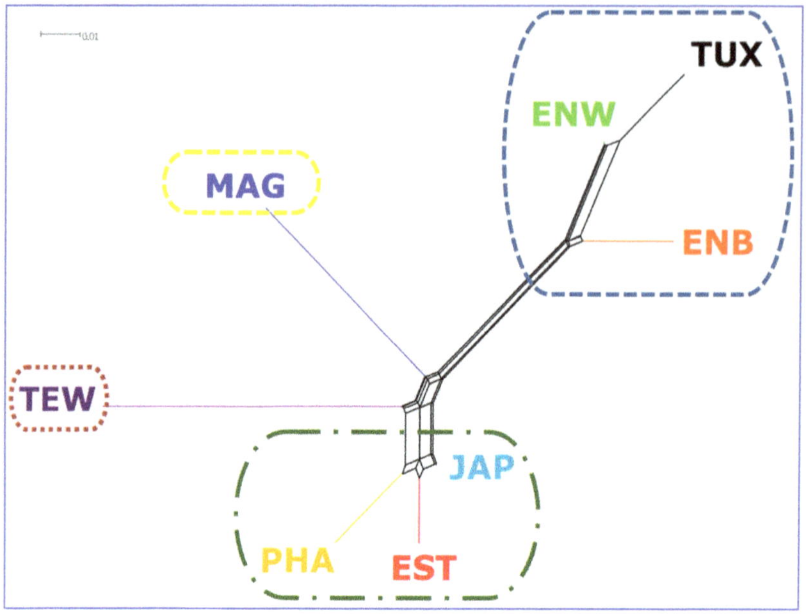

(a)

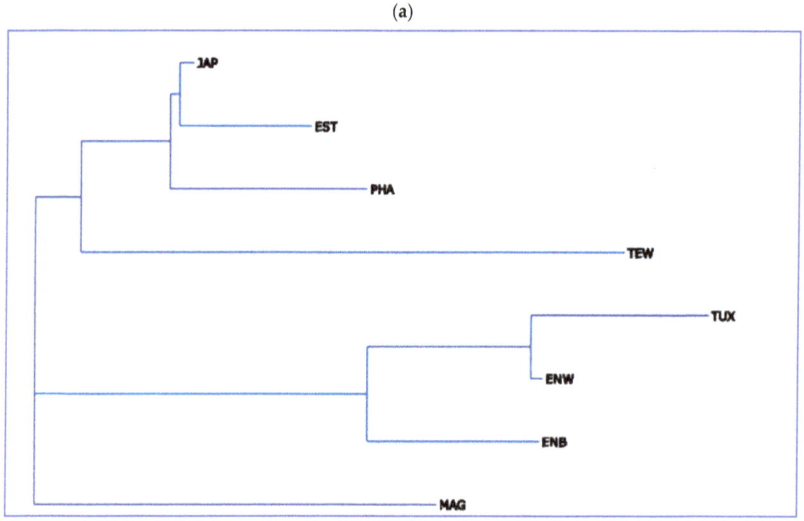

(b)

Figure 5. Phylogenetic trees based on F_{ST} genetic distances characterizing the genetic relationships between the studied quail populations. (**a**) A reconstructed Neighbor-Net network. (**b**) A Neighbor-Joining rootless hierarchical horizontal tree built with proportional edge length and using the Neighbor Joining method [55] and the online T-REX tool [54]. Quail breeds: JAP, Japanese; ENW, English White; ENB, English Black; TUX, Tuxedo; MAG, Manchurian Golden; EST, Estonian; PHA, Pharaoh; TEW, Texas White.

The Neighbor-Net tree based on the values of pairwise FST genetic distances showed that the PHA, EST and JAP populations formed a juncture branch and were located close to each other at the bottom edge of the graph (Figure 5a), indicating their close genetic similarity. The neighboring branch localization of ENW, ENB, as well as TUX on the reconstructed network (Figure 5a) suggested a high genetic similarity of these quail breeds, too. The positioning of the MAG population at the root of the branch suggested that the improvement of this breed occurred mainly due to the selection of purebred quails with a low contribution from other breeds. TEW was also very clearly differentiated from the other breeds, although it was included in one large cluster along with JAP, EST and PHA. The Neighbor-Joining tree had a similar topology (Figure 5b).

Additionally, we analyzed the number and lengths of extended homozygous segments, i.e., ROHs (Table 4, Figure 6).

As can be seen from Figure 6, the greatest numbers of ROHs were within short (0.5–2 Mb) fragments. No ROHs longer than 16 Mb were found in the studied quail breeds. Those longer ROHs may be indicative of recent inbreeding events, and they were discovered in many studies in chickens (e.g., [59–61]). The largest number of fragments of medium length (2–4 and 4–8 Mb) was found in TEW and TUX, suggesting both ongoing breeding work aimed at consolidating desirable traits and the accumulation of homozygous fragments due to the small population size in these breeds. The smallest number and length of ROHs were observed in JAP and PHA.

All variants of migration events obtained using the TreeMix program and different number of iterations (from 1 to 30) are presented in Supplementary Figure S1. The analysis of migration events revealed the presence of expectable gene flows (migrations) between the breeds. In particular, a migration from PHA to TEW was determined for the best number of iterations (29 and 30; mean SE = 0.39; Supplementary Table S3). The respective dendrogram and residual matrix (heat map) are shown in Figure 7. When using other iteration numbers, there were also eight additional observations for the gene flow events between PHA and TEW (Supplementary Figure S1). This was fully confirmed by the known fact of using PHA as one of the progenitor breeds in the creation of TEW. In addition, it can be noted on the other graphs (with different number of iterations; Supplementary Figure S1) that another migration was repeated most often (15 of 30 iteration variants) between ENB and JAP. This also fits perfectly into the origin history of ENB as a mutation of JAP. Four cases of migrations between the ancestor of ENB, ENW and TUX were also observed, which is supported by the facts that ENB and ENW are mutants of JAP and TUX stemmed from crossing ENW and ENB (Table 1).

Table 4. Runs of homozygosity (ROHs) descriptive statistics [1] for the studied breeds.

Breed [2]	ROH Length, Mb (M ± SE)			ROH No. (M ± SE)			F_{ROH} (M ± SE)		
	Average	Min	Max	Average	Min	Max	Average	Min	Max
				Egg type					
JAP	99.04 ± 5.55	48.71	138.58	75.89 ± 3.38	49	104	0.119 ± 0.007	0.06	0.17
ENW	140.00 ± 16.88	8.43	216.10	96.82 ± 10.55	7	133	0.169 ± 0.020	0.01	0.26
ENB	142.92 ± 11.32	58.47	201.15	101.23 ± 6.37	54	137	0.172 ± 0.014	0.07	0.24
TUX	173.54 ± 13.36	36.79	237.00	122.50 ± 7.81	31	164	0.209 ± 0.016	0.04	0.29
MAG	132.63 ± 5.77	91.63	174.35	98.61 ± 3.93	69	121	0.160 ± 0.007	0.11	0.21
				Dual purpose					
EST	114.11 ± 8.23	56.47	137.53	83.11 ± 4.83	49	98	0.137 ± 0.010	0.07	0.17
				Meat type					
PHA	112.18 ± 7.86	62.22	157.96	85.80 ± 5.01	54	108	0.135 ± 0.009	0.07	0.19
TEW	150.66 ± 9.54	115.13	191.32	105.43 ± 4.74	87	122	0.181 ± 0.011	0.14	0.23

[1] ROH No., number of ROHs in a genome; Mb, megabases; M, mean value; SE, standard error; ROH Length, overall length of ROHs in a genome; F_{ROH}, inbreeding coefficient calculated based on ROHs. [2] Quail breeds: JAP, Japanese; ENW, English White; ENB, English Black; TUX, Tuxedo; MAG, Manchurian Golden; EST, Estonian; PHA, Pharaoh; TEW, Texas White.

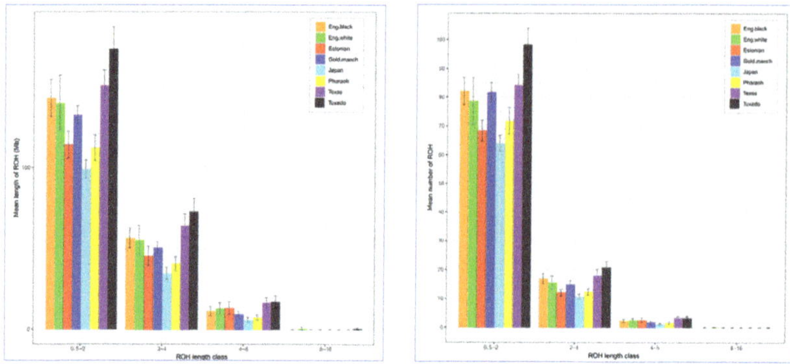

(a) (b)

Figure 6. Descriptive statistics of the runs of homozygosity (ROH) according to ROH length class. (**a**) Overall mean length of ROHs (*Y*-axis) according to ROH length class (*X*-axis; 0.5–2, 2–4, 4–8 and 8–16 Mb) (**b**). Mean number of ROHs (*Y*-axis) according to ROH length class (*X*-axis; 0.5–2, 2–4, 4–8 and 8–16 Mb). Quail breeds: JAP, Japanese; ENW, English White; ENB, English Black; TUX, Tuxedo; MAG, Manchurian Golden; EST, Estonian; PHA, Pharaoh; TEW, Texas White.

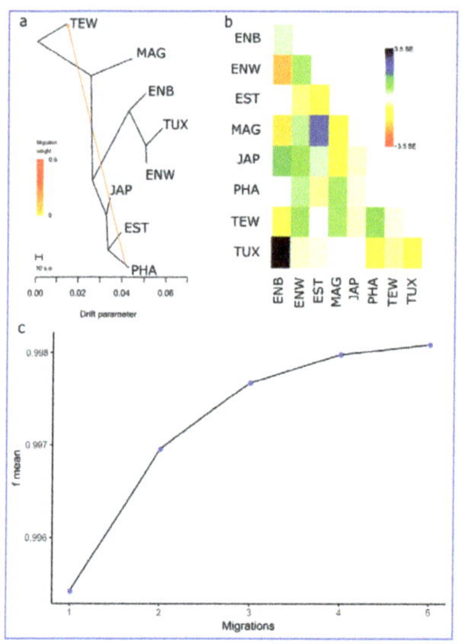

Figure 7. Assessed degree of divergence and the level of gene flow between the studied breeds using 30 iterations. (**a**) Rooted maximum likelihood tree with one migration event. Cut length 10 s.e. corresponds to ten times the average standard error (s.e.) estimated from the sample covariance matrix. Estimated gene flow is shown by an arrow pointing from a donor population (PHA) to a recipient one (TEW) and is colored red in proportion to the intensity of the gene flow. (**b**) Residual matrix derived from the TreeMix analysis for a single migration event expressed as the number of standard error deviations for the observations in the respective breeds. (**c**) Plot representing the proportion of variance (*f*-index) in the sample covariance matrix (¶*W*) accounted for by the model covariance matrix (*W*) as a function of the number of migration events. Quail breeds: JAP, Japanese; ENW, English White; ENB, English Black; TUX, Tuxedo; MAG, Manchurian Golden; EST, Estonian; PHA, Pharaoh; TEW, Texas White.

4. Discussion

In the era of integrative agriculture, there is a need, in the process of monitoring, breeding and selection, to link genetic and genomic technologies to breeding regimes for economically important traits (e.g., [24,62–65]). Performance traits are highest on the list for analysis. Many crucial areas of agricultural production and research such as plant and animal breeding and trait mapping call for reliable and scalable genotyping tools. One such approach that is ideal for non-human organisms is GBS [66–70], which can be effective for integrating genomic and performance data. In this regard, we made an attempt to demonstrate how "congruent" interbreed patterns of genomic architecture are with those for productivity traits in quails. In our GBS study, we, for the first time, collated and compared eight breeds of domestic quail. These represent a large share (~11%) of the world gene pool of quail breeds and three purposes of their use (in terms of productive traits), and they also illustrate the evolutionary component of the selection of individuals in the process of domestication and breeding of this bird species. Having assessed the performance traits and using IPI, we quite accurately confirmed the initial (conventional) classification of these breeds that has been established in the quail breeding practice depending on their selection direction and utility type.

The revealed phylogeny pattern based on genomic data, however, had a lower congruence with the breed configuration obtained from productivity traits using IPI. We note that the egg-type breeds TUX, ENB and ENW, which form a single cluster according to genomic data (see, for example, Figure 5), are located in Table 2 on three adjacent rows (with IPI from 8.3 to 9.0). Similarly, it can be seen that the meat-type breeds PHA and TEW, also located on two adjacent rows in Table 2 (IPI = ~5), were included in one large cluster on the phylogenetic tree. No other similar patterns were observed. Apparently, genomic data reflect not only the selection direction and utility type (due to specific breeding work with breeds), but also other features of the breeds, e.g., the history of their development (namely, the original breeds and populations), as well as genetic processes occurring in individual populations (gene flow, genetic drift, random or purposeful crossbreeding, etc.). Taking into account all of the above, we can confirm that, in a general sense and to a certain degree, data on history, management and phenotype are congruent with the description of the diversity between breeds/populations (e.g., [71]).

When analyzing genetic diversity of the eight breeds, we observed the difference between inbreeding measures based on F_{IS} and ROH metrics. Most likely, these different estimates were a consequence of different approaches to calculating the two inbreeding indices. The F_{IS} score is calculated based on the observed and expected heterozygosities and reflects a lack (or excess) of heterozygotes, while F_{ROH} conforms to the proportion of homozygous regions in the genome. The latter estimate appears to be more accurate than the former one because it directly assesses genomic homozygosity. These two indicators are normally calculated and reported in similar studies (e.g., [59,61,70]) to describe different aspects of the defined genomic diversity and homozygosity.

To date, there have been a number of investigations focusing on the genetic diversity in quails. These studies were mainly related to the assessment of the genotypes of domestic and wild quails in order to characterize the genetic structure of populations in these species [22,72,73], as well as the issues of their hybridization in the wild [74–76]. Most of the work in this area was executed using microsatellite and mtDNA markers [72–77]. However, the use of conventional molecular markers has drawbacks and limitations in the case of both mtDNA (e.g., [78–80]) and microsatellite markers (e.g., [81,82]). SNP-assisted applications are more advantageous (e.g., [83,84], and as demonstrated herein). In this paper, to explore the biology of different quail species, we employed a relatively new approach through the use of SNPs obtained via whole-genome sequencing and subsequent GBS analysis. The relevance and efficiency of implementing GBS for genome-wide genotyping has also been demonstrated in other poultry, model and non-model, species, including chickens [85], ducks [86,87] and geese [88]. For instance, Grzegorczyk et al. [88] studied the genetic diversity and phylogenetic relationships of 12 Polish goose breeds using the

GBS approach and identified SNPs associated with economic traits. Zhu et al. [86] developed and tested the GBS protocol for ducks, which resulted in 169,209 significant SNPs. Using GBS in ducks, SNPs and genes associated with plumage color [87] and 18 carcass traits [89] were also identified.

This approach can sometimes be employed in combination with traditional molecular markers. For example, Mathur and DeWoody [90] examined the genetic diversity of three populations of the wild Montezuma quail (*Cyrtonyx montezumae*) using whole-genome sequencing data from 74 quails. Ogada et al. [22] utilized both mtDNA and GBS analyses to investigate the genetic diversity and demographic history of wild African harlequin quail populations of Siaya County. In the current study, we demonstrated the effectiveness of SNPs, as the currently most used molecular markers, and the GBS approach for a genome-wide comparative assessment of different breeds of domestic quail. At the same time, we confirmed the high resolution and analytical power of the GBS-derived SNP scanning method for solving problems in modern genetic research, as has been previously shown in similar studies [83,91]. For instance, Weigend et al. [83] reported the evaluation of clustering accuracy for 10 chicken populations into 10 cluster groups based on microsatellite markers and SNPs. Given that the SNP data generally contained more alleles than microsatellites, these two sets of data allowed a comparison between microsatellites and SNPs as genetic markers for biodiversity research in favor of more informative genome-wide SNPs.

Using GBS analysis for the first time to analyze the genomes of different domestic quail breeds, we observed a lower genetic diversity in TUX as compared to JAP quails ($_{U}H_{E}$ = 0.265 vs. 0.319; Table 3). A possible reason for this may be genetic drift in the TUX population that is small in size (13 animals) and has been bred for a long time as a closed population. At the same time, the higher level of genetic diversity in JAP may also reflect the crossbred origin of the individuals used in this study. Our study confirmed that GBS analysis can be considered an appropriate tool for investigating intraspecific differentiation. Therefore, the discovered similarities/differences can be used as a marker of gene flow among the studied breed samples as was shown by us here as a result of TreeMix-assisted analysis of migration edges.

The results of PCA plotting, Neighbor-Net and Admixture clustering and other related genomic analyses (Figures 3–7) clearly distinguished quail breeds and utility types in line with their specific genetic origins and selection for economically important traits. This differentiation can provide important information for the collection, conservation, research and utilization of quail genetic resources as shown in other poultry breeds [26,92–96]. In particular, using two genetically divergent breeds, JAP and TEW (e.g., Figures 2, 3 and 5), we recently created a model resource F_2 population to perform a GWAS analysis of growth dynamics in quails [97]. As a result of crossing these two contrasting breeds (slow-growing egg-type JAP and fast-growing meat-type TEW), the F_2 population had a significant range of variability in BW and other phenotypic traits and was instrumental in identifying a series of SNPs associated with BW and a number of the respective candidate genes [97].

Collectively, based on SNP genotypes using GBS analysis, our findings illustrated phylogenetic relationships for the eight quail breeds that represented the egg type, meat type and dual purpose of use. The phylogenetic trees built on the basis of GBS data showed that the JAP, PHA and EST breeds were genetically similar to a certain degree. In addition, according to the obtained tree configuration, it can be argued that these three breeds can be especially valuable sources of genetic variability since they were close to the root of the phylogenetic tree. MAG and TEW can be classified as genetically more distant relative to the other breeds studied and to one another. The clustering of ENB, ENW and TUX into one group corresponded, in terms of their relatedness, to the historical records of the development of these breeds. Overall, we were able to show that GBS analysis is an efficient and useful instrument for elucidating genomic architecture and divergence across different quail breeds.

5. Conclusions

Using the GBS molecular method in the present investigation, we evaluated, for the first time, the genetic diversity of the eight quail breeds (representing about 11% of the global quail germplasm) and identified their evolutionary relationships suggesting a possible relation to their performance. In particular, the respective genetic divergence was shown for the egg (JAP, ENW, ENB, TUX and MAG), meat (TEW and PHA) and dual-purpose (EST) utility types. This study contributes to the identification of genetic differentiation and determination of relatedness between the studied breeds. In addition, it was demonstrated for the first time that the GBS analysis method is instrumental in the intra- and inter-breed assessment of genetic variation in domestic quails. The information reported here facilitates a deeper understanding of the processes of breed formation and selection in quails and can be further used to improve their economically important traits by identifying significant SNPs and candidate genes associated with these traits.

Supplementary Materials: The following supporting information can be downloaded at: https://www.mdpi.com/article/10.3390/ani13223439/s1, Figure S1: Assessment of the degree of divergence and the level of gene flow between the studied quail breeds; Table S1: Pairwise interbreed Euclidean distances obtained for breed IPI values using the Phantasus program; Table S2: Pairwise F_{ST} based interbreed genetic distances; Table S3: Mean SE values according to iteration number when calculating migration events obtained using the TreeMix program.

Author Contributions: Conceptualization, N.A.V. and N.A.Z.; methodology, A.S.A., P.V.L., D.V.A., V.I.F., V.G.N., J.S., G.B., J.C.M. and R.B.; software, A.S.A., P.V.L. and A.V.S.; validation, N.A.V., N.Y.G., A.N.V. and L.A.V.; formal analysis, M.N.R., A.S.A., P.V.L., A.V.S., D.V.A., V.I.F., V.G.N., J.S., G.B., J.C.M. and R.B.; investigation, N.Y.G., A.N.V., L.A.V., J.C.M. and R.B.; resources, D.V.A. and V.I.F.; data curation, N.A.V.; writing—original draft preparation, N.A.V., M.N.R. and P.V.L.; writing—review and editing, N.A.V., M.N.R., D.K.G., J.S., J.C.M. and N.A.Z.; visualization, N.A.V., M.N.R., A.S.A. and A.V.S.; supervision, N.A.V., D.K.G. and N.A.Z.; project administration, N.A.V. and N.A.Z.; funding acquisition, N.A.V. and N.A.Z. All authors have read and agreed to the published version of the manuscript.

Funding: This work was funded by the Russian Science Foundation, Grant No. 21-16-00086 (https://rscf.ru/en/project/21-16-00086, accessed on 25 September 2023), and by the Ministry of Science and Higher Education of the Russian Federation, Grant No. 075-15-2021-1037 (Internal No. 15.BRK.21.0001).

Institutional Review Board Statement: The study was conducted according to the guidelines of the Declaration of Helsinki and the LKEFRCAH ethical guidelines. Protocol No. 7 was approved by the LKEFRCAH Commission on the BioEthics of Animal Experiments on 10 August 2021

Informed Consent Statement: Not applicable.

Data Availability Statement: The sequence data is accessible to readers on request.

Acknowledgments: Genotyping-by-sequencing services were provided by AgResearch (Lincoln, Canterbury, New Zealand; https://www.agresearch.co.nz/partnering-with-us/products-and-services/genomnz/genotyping-services/, accessed on 25 September 2023), under curation of Tracey Van Stijn (tracey.vanstijn@agresearch.co.nz).

Conflicts of Interest: The authors declare no conflict of interest.

References

1. Venkitanarayanan, K.; Thakur, S.; Ricke, S.C. (Eds.) *Food Safety in Poultry Meat Production*; Springer International Publishing: Cham, Switzerland, 2019; ISBN 978-3-030-05010-8/978-3-030-05011-5. [CrossRef]
2. Vorotnikov, V. *Russia Puts Focus on 'Non-Traditional' Poultry*; FoodNavigator Europe; William Reed Ltd.: Crawley, UK, 2014. Available online: https://www.foodnavigator.com/Article/2014/01/16/Russia-s-non-traditional-poultry-sees-planned-boost (accessed on 25 September 2023).
3. Minvielle, F. What are quail good for in a chicken-focused world? *Worlds Poult. Sci. J.* **2009**, *65*, 601–608. [CrossRef]

4. Romanov, M.N.; Sazanov, A.A.; Moiseyeva, I.G.; Smirnov, A.F. Poultry. In *Genome Mapping and Genomics in Animals, Vol. 3: Genome Mapping and Genomics in Domestic Animals*; Cockett, N.E., Kole, C., Eds.; Springer: Berlin/Heidelberg, Gemany; New York, NY, USA, 2009; pp. 75–141. ISBN 978-3-540-73834-3/978-3-540-73835-0. [CrossRef]
5. Podstreshnyi, O.P.; Tereshchenko, O.V.; Katerynych, O.O.; Tkachyk, T.E.; Podstreshna, I.O. *Production of Quail Eggs and Meat: Methodical Recommendations*, 2nd ed.; Tereshchenko, O.V., Ed.; Poultry Research Institute, NAAS of Ukraine: Birky, Ukraine, 2010. Available online: https://www.researchgate.net/publication/342802513_Podstresnij_OP_Teresenko_OV_Katerinic_OO_Tkacik_TE_Podstresna_IO_Virobnictvo_perepelinih_aec_ta_m\T1\textquoterightasa_metodicni_rekomendacii_Uklad_OV_Teresenko_ta_in_pid_red_OV_Teresenka_-_2-e_vid_pererob_ta_dop_-_Bi (accessed on 25 September 2023). (In Ukrainian)
6. Podstreshnyi, O.; Tereshchenko, O. Maintenance of adult quails. *Ahrar. Krayina* **2012**, *6*, 8–9. Available online: https://www.researchgate.net/publication/342832587_Podstresnij_O_Teresenko_O_Utrimanna_doroslih_perepeliv_Agrarna_kraina_-_2012_-_Cerven-_S_8-9 (accessed on 25 September 2023). (In Ukrainian).
7. Podstreshnyi, O.; Tereshchenko, O. Feeding young quails. *Ahrar. Krayina* **2012**, *7*, 6. Available online: https://www.researchgate.net/publication/342832583_Podstresnij_O_Teresenko_O_Godivla_molodnaka_perepeliv_Agrarna_kraina_-_2012_-_Lipen_-_S_6_httpagrokrainacomuapoultry_farming268-godvlya-molodnyaka-perepelvhtml (accessed on 25 September 2023). (In Ukrainian).
8. Shimakura, K. Notes on the genetics of the Japanese quail: I. The simple, Mendelian, autosomal, recessive character, "brown-splashed white", of its plumage. *Jpn. J. Genet.* **1940**, *16*, 106–112. (In Japanese with English Summary). [CrossRef]
9. Huss, D.; Poynter, G.; Lansford, R. Japanese quail (*Coturnix japonica*) as a laboratory animal model. *Lab. Anim.* **2008**, *37*, 513–519. [CrossRef]
10. Chang, G.B.; Chang, H.; Liu, X.P.; Xu, W.; Wang, H.Y.; Zhao, W.M.; Olowofeso, O. Developmental research on the origin and phylogeny of quails. *Worlds Poult. Sci. J.* **2005**, *61*, 105–112. [CrossRef]
11. Romanov, M.N.; Wezyk, S.; Cywa-Benko, K.; Sakhatsky, N.I. Poultry genetic resources in the countries of Eastern Europe—History and current state. *Poult. Avian Biol. Rev.* **1996**, *7*, 1–29. Available online: https://www.researchgate.net/publication/255710929_Poultry_genetic_resources_in_the_countries_of_Eastern_Europe_-_history_and_current_state (accessed on 25 September 2023).
12. Volkovoy, S.; Bondarenko, Y. Japanese quail plumage rainbow. *Priusadebnoye Khozyaystvo* **1989**, *5*, 14–15. Available online: https://yablonka.net/world/zh/686-raduga-opereniya-yaponskogo-perepela.html (accessed on 25 September 2023). (In Russian).
13. Genofond. *Catalogue of Breeds: Quails*; Official Site of the Company Genofond LLC: Sergiev Posad, Russia, 2015. Available online: http://www.genofond-sp.ru/quail.html (accessed on 25 September 2023). (In Russian)
14. Baumgartner, J.; Bondarenko, Y.V. Search for Autosexing Strains and Crosses in Japanese Quail. In Proceedings of the 8th International Symposium on Actual Problems of Avian Genetics, Smolenice, Czechoslovakia, 3–6 April 1989; Slovak Society for Agriculture, Forestry, Food and Veterinary Sciences of Slovak Academy of Sciences: Bratislava, Czechoslovakia; Poultry Research and Production Institute: Ivanka pri Dunaji, Bratislava, Czechoslovakia; Czechoslovak Branch of WPSA: Smolenice, Czechoslovakia, 1989; pp. 262–265.
15. Bondarenko, Y.V. *Contemporary Methods for Determining the Sex of Young Domestic and Ornamental Birds*, 4th ed.; NTUL: Sumy, Ukraine, 2020. Available online: https://web.archive.org/web/20220325163425/https:/repo.snau.edu.ua/bitstream/123456789/8319/1/%D0%9A%D0%BD%D0%B8%D0%B3%D0%B0%20%D0%91%D0%BE%D0%BD%D0%B4%D0%B0%D1%80%D0%B5%D0%BD%D0%BA%D0%BE_%2B17.12.%202019%20%D0%A0%D0%BE%D0%B7%D0%B1%D0%BB%D0%BE%D0%BA%D0%BE%D0%B2%D0%B0%D0%BD%D0%B0.pdf (accessed on 25 September 2023). (In Russian)
16. Elshire, R.J.; Glaubitz, J.C.; Sun, Q.; Poland, J.A.; Kawamoto, K.; Buckler, E.S.; Mitchell, S.E. A robust, simple genotyping-by-sequencing (GBS) approach for high diversity species. *PLoS ONE* **2011**, *6*, e19379. [CrossRef]
17. Gurgul, A.; Miksza-Cybulska, A.; Szmatoła, T.; Jasielczuk, I.; Piestrzyńska-Kajtoch, A.; Fornal, A.; Semik-Gurgul, E.; Bugno-Poniewierska, M. Genotyping-by-sequencing performance in selected livestock species. *Genomics* **2019**, *111*, 186–195. [CrossRef] [PubMed]
18. Sonah, H.; Bastien, M.; Iquira, E.; Tardivel, A.; Légaré, G.; Boyle, B.; Normandeau, É.; Laroche, J.; Larose, S.; Jean, M.; et al. An improved genotyping by sequencing (GBS) approach offering increased versatility and efficiency of SNP discovery and genotyping. *PLoS ONE* **2013**, *8*, e54603. [CrossRef] [PubMed]
19. Jacobs, J.; Clarke, S.; Faville, M.; Griffiths, A.; Cao, M.; Tan, R.; Van Stijn, T.; Anderson, R.; Ashby, R.; Rowe, S.; et al. Genotyping-by-sequencing Applications in Biology. In Proceedings of the Plant and Animal Genome XXV Conference, San Diego, CA, USA, 13–18 January 2017; Scherago International: Surfside, FL, USA, 2017. Abstract P0128. [CrossRef]
20. De Donato, M.; Peters, S.O.; Mitchell, S.E.; Hussain, T.; Imumorin, I.G. Genotyping-by-sequencing (GBS): A novel, efficient and cost-effective genotyping method for cattle using next-generation sequencing. *PLoS ONE* **2013**, *8*, e62137. [CrossRef] [PubMed]
21. Larson, W.A.; Seeb, L.W.; Everett, M.V.; Waples, R.K.; Templin, W.D.; Seeb, J.E. Genotyping by sequencing resolves shallow population structure to inform conservation of Chinook salmon (*Oncorhynchus tshawytscha*). *Evol. Appl.* **2014**, *7*, 355–369. [CrossRef] [PubMed]
22. Ogada, S.; Otecko, N.O.; Moraa Kennedy, G.; Musina, J.; Agwanda, B.; Obanda, V.; Lichoti, J.; Peng, M.S.; Ommeh, S. Demographic history and genetic diversity of wild African harlequin quail (*Coturnix delegorguei delegorguei*) populations of Kenya. *Ecol. Evol.* **2021**, *11*, 18562–18574. [CrossRef] [PubMed]

23. Ravagni, S.; Sanchez-Donoso, I.; Jiménez-Blasco, I.; Andrade, P.; Puigcerver, M.; Chorão Guedes, A.; Godinho, R.; Gonçalves, D.; Leitão, M.; Leonard, J.A.; et al. Evolutionary history of an island endemic, the Azorean common quail. *Mol. Ecol.* **2023**. [CrossRef] [PubMed]
24. Prituzhalova, A.O.; Volkova, N.A.; Kuzmina, T.I.; Vetokh, A.N.; Dzhagaev, A.Y. Monitoring of indicators of chromatin status in quails ovarian follicles granulosa cells of different directions of productivity. *Agrar. Nauka* **2023**, *368*, 53–57. (In Russian with English Summary). [CrossRef]
25. Mills, A.D.; Crawford, L.L.; Domjan, M.; Faure, J.M. The behavior of the Japanese or domestic quail *Coturnix japonica*. *Neurosci. Biobehav. Rev.* **1997**, *21*, 261–281. [CrossRef] [PubMed]
26. Ryabokon, Y.O.; Pabat, V.O.; Mykytyuk, D.M.; Frolov, V.V.; Katerynych, O.O.; Bondarenko, Y.V.; Mosyakina, T.V.; Gadyuchko, O.T.; Kovalenko, G.T.; Gritsenko, D.M.; et al. *Catalog of Poultry Breeding Resources of Ukraine*; Ryabokon, Y.O., Ed.; Poultry Research Institute: Kharkiv, Ukraine, 2005. Available online: http://avianua.com/archiv/plevreestr/per.pdf (accessed on 25 September 2023). (In Ukrainian)
27. Domesticfutures. Quail Breeds: Characteristics with Photos; domesticfutures.com. 2021. Available online: https://domesticfutures.com/porody-perepelov-harakteristiki-s-fotografiyami-4457 (accessed on 25 September 2023).
28. Genchev, A. Egg production potential of Manchurian Golden quail breeders. *Agric. Sci. Technol.* **2011**, *3*, 73–80. Available online: http://agriscitech.eu/wp-content/uploads/2014/05/GB_02.pdf (accessed on 25 September 2023).
29. Purely Poultry. *Gold Coturnix Quail Set*; Purely Poultry: Durand, WI, USA, 2023. Available online: https://www.purelypoultry.com/index.php?main_page=product_info&products_id=1267 (accessed on 25 September 2023).
30. German, N.Y.; Vetokh, A.N.; Dzhagaev, A.Y.; Ilyina, E.R.; Kotova, T.O. Morphometric parameters of eggs from breeds quail for meat. *Vet. Kormlenie* **2023**, *2*, 20–23. (In Russian with English Summary). [CrossRef]
31. Vakhrameev, A.B.; Narushin, V.G.; Larkina, T.A.; Barkova, O.Y.; Peglivanyan, G.K.; Dysin, A.P.; Dementieva, N.V.; Makarova, A.V.; Shcherbakov, Y.S.; Pozovnikova, M.V.; et al. Disentangling clustering configuration intricacies for divergently selected chicken breeds. *Sci. Rep.* **2023**, *13*, 3319. [CrossRef] [PubMed]
32. GraphPad Software. Dotmatics. Available online: https://www.graphpad.com/ (accessed on 18 October 2023).
33. Dodds, K.G.; McEwan, J.C.; Brauning, R.; Anderson, R.M.; van Stijn, T.C.; Kristjánsson, T.; Clarke, S.M. Construction of relatedness matrices using genotyping-by-sequencing data. *BMC Genom.* **2015**, *16*, 1047. [CrossRef] [PubMed]
34. AgResearch. DECONVQC; GitHub, Inc. 2016. Available online: https://github.com/AgResearch/DECONVQC (accessed on 25 September 2023).
35. Andrews, S. *FastQC: A Quality Control Tool for High Throughput Sequence Data*; Version 0.10.1; Bioinformatics Group, Babraham Institute: Cambridge, UK, 2012. Available online: http://www.bioinformatics.babraham.ac.uk/projects/fastqc (accessed on 25 September 2023).
36. Morris, K.M.; Hindle, M.M.; Boitard, S.; Burt, D.W.; Danner, A.F.; Eory, L.; Forrest, H.L.; Gourichon, D.; Gros, J.; Hillier, L.W.; et al. The quail genome: Insights into social behaviour, seasonal biology and infectious disease response. *BMC Biol.* **2020**, *18*, 14. [CrossRef]
37. Szpak, M. Ensembl 104 Has Been Released; Ensembl Blog. 2021. Available online: https://www.ensembl.info/2021/05/05/ensembl-104-has-been-released/ (accessed on 25 September 2023).
38. Martin, M. Cutadapt removes adapter sequences from high-throughput sequencing reads. *EMBnet J.* **2011**, *17*, 10–12. [CrossRef]
39. Martin, M. Cutadapt. Version 3.4; GitHub, Inc. 2021. Available online: https://github.com/marcelm/cutadapt (accessed on 25 September 2023).
40. Kang, J.; Dodds, K.; Byrne, S.; Faville, M.; Black, M.; Hess, A.; Hess, M.; McCulloch, A.; Jacobs, J.; Milbourne, D.; et al. snpGBS: A Simple and Flexible Bioinformatics Workflow to Identify SNPs from Genotyping-by-sequencing Data. In *Exploiting Genetic Diversity of Forages to Fulfil Their Economic and Environmental Roles, Proceedings of the 34th Meeting of the EUCARPIA Fodder Crops and Amenity Grasses Section in Cooperation with the EUCARPIA Festulolium Working Group, Freising, Germany, 6–8 September 2021*; Hartmann, S., Bachmann-Pfabe, S., Byrne, S., Feuerstein, U., Julier, B., Kölliker, R., Kopecky, D., Roldan-Ruiz, I., Ruttink, T., Sampoux, J.-P., et al., Eds.; Palacký University Press: Olomouc, Czech Republic, 2021; pp. 67–70. ISBN 978-80-244-5967-7/978-80-244-5968-4/978-80-244-5969-1. [CrossRef]
41. AgResearch. snpGBS; GitHub, Inc. 2021. Available online: https://github.com/AgResearch/snpGBS (accessed on 25 September 2023).
42. Langmead, B. bowtie2: A Fast and Sensitive Gapped Read Aligner. Version 2.4.4; GitHub, Inc. 2021. Available online: https://github.com/BenLangmead/bowtie2 (accessed on 25 September 2023).
43. Li, H.; Handsaker, B.; Wysoker, A.; Fennell, T.; Ruan, J.; Homer, N.; Marth, G.; Abecasis, G.; Durbin, R.; 1000 Genome Project Data Processing Subgroup. The Sequence Alignment/Map format and SAMtools. *Bioinformatics* **2009**, *25*, 2078–2079. [CrossRef] [PubMed]
44. Danecek, P.; Bonfield, J.K.; Liddle, J.; Marshall, J.; Ohan, V.; Pollard, M.O.; Whitwham, A.; Keane, T.; McCarthy, S.A.; Davies, R.M.; et al. Twelve years of SAMtools and BCFtools. *GigaScience* **2021**, *10*, giab008. [CrossRef]
45. R Core Team. *R: A Language and Environment for Statistical Computing*; R Foundation for Statistical Computing: Vienna, Austria, 2018. Available online: https://www.r-project.org/ (accessed on 25 September 2023).
46. Purcell, S.; Neale, B.; Todd-Brown, K.; Thomas, L.; Ferreira, M.A.; Bender, D.; Maller, J.; Sklar, P.; de Bakker, P.I.; Daly, M.J.; et al. PLINK: A tool set for whole-genome association and population-based linkage analyses. *Am. J. Hum. Genet.* **2007**, *81*, 559–575. [CrossRef]

47. Keenan, K.; McGinnity, P.; Cross, T.F.; Crozier, W.W.; Prodohl, P.A. diveRsity: An R package for the estimation and exploration of population genetics parameters and their associated errors. *Methods Ecol. Evol.* **2013**, *4*, 782–788. [CrossRef]
48. Kalinowski, S.T. Counting alleles with rarefaction: Private alleles and hierarchical sampling designs. *Conserv. Genet.* **2004**, *5*, 539–543. [CrossRef]
49. Wickham, H. *ggplot2: Elegant Graphics for Data Analysis*; Springer: New York, NY, USA, 2009; ISBN 978-0-387-98141-3. [CrossRef]
50. Huson, D.H.; Bryant, D. Application of phylogenetic networks in evolutionary studies. *Mol. Biol. Evol.* **2006**, *23*, 254–267. [CrossRef]
51. Alexander, D.H.; Novembre, J.; Lange, K. Fast model-based estimation of ancestry in unrelated individuals. *Genome Res.* **2009**, *19*, 1655–1664. [CrossRef]
52. Milanesi, M.; Capomaccio, S.; Vajana, E.; Bomba, L.; Garcia, J.F.; Ajmone-Marsan, P.; Colli, L. BITE: An R package for biodiversity analyses. *bioRxiv* **2017**, 181610. [CrossRef]
53. Zenkova, D.; Kamenev, V.; Sablina, R.; Artyomov, M.; Sergushichev, A. Phantasus: Visual and Interactive Gene Expression Analysis. 2018. Available online: https://ctlab.itmo.ru/phantasus (accessed on 25 September 2023). [CrossRef]
54. Boc, A.; Diallo, A.B.; Makarenkov, V. T-REX: A web server for inferring, validating and visualizing phylogenetic trees and networks. *Nucleic Acids Res.* **2012**, *40*, W573–W579. [CrossRef] [PubMed]
55. Saitou, N.; Nei, M. The neighbor-joining method: A new method for reconstructing phylogenetic trees. *Mol. Biol. Evol.* **1987**, *4*, 406–425. [CrossRef]
56. Pickrell, J.K.; Pritchard, J.K. Inference of population splits and mixtures from genome-wide allele frequency data. *PLoS Genet.* **2012**, *8*, e1002967. [CrossRef] [PubMed]
57. Fitak, R.R. *OptM*: Estimating the optimal number of migration edges on population trees using *Treemix*. *Biol. Methods Protoc.* **2021**, *6*, bpab017. [CrossRef] [PubMed]
58. Biscarini, F.; Paolo Cozzi, P.; Gaspa, G.; Marras, G. *detectRUNS: Detect Runs of Homozygosity and Runs of Heterozygosity in Diploid Genomes*; R Package Version 0.9.6; The Comprehensive R Archive Network (CRAN); Institute for Statistics and Mathematics, Vienna University of Economics and Business: Vienna, Austria, 2019. Available online: https://CRAN.R-project.org/package=detectRUNS (accessed on 25 September 2023).
59. Abdelmanova, A.S.; Dotsev, A.V.; Romanov, M.N.; Stanishevskaya, O.I.; Gladyr, E.A.; Rodionov, A.N.; Vetokh, A.N.; Volkova, N.A.; Fedorova, E.S.; Gusev, I.V.; et al. Unveiling comparative genomic trajectories of selection and key candidate genes in egg-type Russian White and meat-type White Cornish chickens. *Biology* **2021**, *10*, 876. [CrossRef]
60. Rostamzadeh Mahdabi, E.; Esmailizadeh, A.; Ayatollahi Mehrgardi, A.; Asadi Fozi, M. A genome-wide scan to identify signatures of selection in two Iranian indigenous chicken ecotypes. *Genet. Sel. Evol.* **2021**, *53*, 72. [CrossRef]
61. Cendron, F.; Mastrangelo, S.; Tolone, M.; Perini, F.; Lasagna, E.; Cassandro, M. Genome-wide analysis reveals the patterns of genetic diversity and population structure of 8 Italian local chicken breeds. *Poult. Sci.* **2021**, *100*, 441–451. [CrossRef]
62. Moiseyeva, I.G.; Bannikova, L.V.; Altukhov, Y.P. State of poultry breeding in Russia: Genetic monitoring. *Mezhdunar. S-kh. Zh.* **1993**, *5–6*, 66–69. (In Russian)
63. Bondarenko, Y.V.; Kutnyuk, P.I. Some Results of Genetic Monitoring of Embryonic Defects in Poultry Populations. In *Gene Pool of Animal Breeds and Methods of its Use, Proceedings of the Materials of the International Scientific and Practical Conference Dedicated to the 110th Anniversary of the Birth of Academician N.D. Potemkin, Kharkov, Ukraine, 5–6 December 1995*; Ministry of Agriculture and Food of Ukraine, Kharkov Zooveterinary Institute, RIO KhZVI: Kharkov, Ukraine, 1995; pp. 63–64. (In Russian)
64. Bondarenko, Y.V.; Podstreshny, A.P. Genetic Monitoring of Chicken Populations. In Proceedings of the 2nd International Conference on Molecular Genetic Markers of Animals, Kiev, Ukraine, 15–17 May 1996; Agrarna Nauka: Kiev, Ukraine, 1996; pp. 47–48. (In Russian).
65. Zakharov-Gesekhus, I.A.; Stolpovsky, Y.A.; Ukhanov, S.V.; Moiseyeva, I.G.; Sulimova, G.E. Monitoring the gene pools of animal populations in connection with selection tasks and the study of phylogeny. In *Farm Animals*; Russian Academy of Sciences: Moscow, Russia, 2007; pp. 122–124. (In Russian)
66. Heffelfinger, C.; Fragoso, C.A.; Moreno, M.A.; Overton, J.D.; Mottinger, J.P.; Zhao, H.; Tohme, J.; Dellaporta, S.L. Flexible and scalable genotyping-by-sequencing strategies for population studies. *BMC Genom.* **2014**, *15*, 979. [CrossRef] [PubMed]
67. Mischler, C.; Veale, A.; Van Stijn, T.; Brauning, R.; McEwan, J.C.; Maloney, R.; Robertson, B.C. Population connectivity and traces of mitochondrial introgression in New Zealand black-billed gulls (*Larus bulleri*). *Genes* **2018**, *9*, 544. [CrossRef] [PubMed]
68. Rexer-Huber, K.; Veale, A.J.; Catry, P.; Cherel, Y.; Dutoit, L.; Foster, Y.; McEwan, J.C.; Parker, G.C.; Phillips, R.A.; Ryan, P.G.; et al. Genomics detects population structure within and between ocean basins in a circumpolar seabird: The white-chinned petrel. *Mol. Ecol.* **2019**, *28*, 4552–4572. [CrossRef]
69. Wold, J.R.; Robertson, C.J.; Chambers, G.K.; Van Stijn, T.; Ritchie, P.A. Genetic connectivity in allopatric seabirds: Lack of inferred gene flow between Northern and Southern Buller's albatross populations (*Thalassarche bulleri* ssp.). *Emu-Austral Ornithol.* **2021**, *121*, 113–123. [CrossRef]
70. Foster, Y.; Dutoit, L.; Grosser, S.; Dussex, N.; Foster, B.J.; Dodds, K.G.; Brauning, R.; Van Stijn, T.; Robertson, F.; McEwan, J.C.; et al. Genomic signatures of inbreeding in a critically endangered parrot, the kākāpō. *G3* **2021**, *11*, jkab307. [CrossRef] [PubMed]
71. Tixier-Boichard, M.; Coquerelle, G.; Vilela-Lamego, C.; Weigend, S.; Barre-Dirrie, A.; Groenen, M.; Crooijmans, R.; Vignal, A.; Hillel, J.; Freidlin, P.; et al. Contribution of Data on History, Management and Phenotype to the Description of the Diversity between Chicken Populations Sampled within the AVIANDIV Project. In Proceedings of the Poultry Genetics Symposium,

Mariensee, Germany, 6–8 October 1999; Preisinger, R., Ed.; Working Group 3 of WPSA, Lohmann Tierzucht: Cuxhaven, Germany, 1999; pp. 15–21. Available online: https://jukuri.luke.fi/handle/10024/446389 (accessed on 25 September 2023).
72. Nunome, M.; Nakano, M.; Tadano, R.; Kawahara-Miki, R.; Kono, T.; Takahashi, S.; Kawashima, T.; Fujiwara, A.; Nirasawa, K.; Mizutani, M.; et al. Genetic divergence in domestic Japanese quail inferred from mitochondrial DNA D-loop and microsatellite markers. *PLoS ONE* **2017**, *12*, e0169978. [CrossRef]
73. Smith, S.; Fusani, L.; Boglarka, B.; Sanchez-Donoso, I.; Marasco, V. Lack of introgression of Japanese quail in a captive population of common quail. *Eur. J. Wildl. Res.* **2018**, *64*, 51. [CrossRef]
74. Amaral, A.J.; Silva, A.B.; Grosso, A.R.; Chikhi, L.; Bastos-Silveira, C.; Dias, D. Detection of hybridization and species identification in domesticated and wild quails using genetic markers. *Folia Zool.* **2007**, *56*, 285–300. Available online: https://www.ivb.cz/wp-content/uploads/56_285-300.pdf (accessed on 25 September 2023).
75. Barilani, M.; Deregnaucourt, S.; Gallego, S.; Galli, L.; Mucci, N.; Piombo, R.; Puigcerver, M.; Rimondi, S.; Rodríguez-Teijeiro, J.D.; Spanò, S.; et al. Detecting hybridization in wild (*Coturnix c. coturnix*) and domesticated (*Coturnix c. japonica*) quail populations. *Biol. Conserv.* **2005**, *126*, 445–455. [CrossRef]
76. Chazara, O.; Minvielle, F.; Roux, D.; Bed'hom, B.; Feve, K.; Coville, J.-L.; Kayang, B.B.; Lumineau, S.; Vignal, A.; Boutin, J.-M.; et al. Evidence for introgressive hybridization of wild common quail (*Coturnix coturnix*) by domesticated Japanese quail (*Coturnix japonica*) in France. *Conserv. Genet.* **2010**, *11*, 1051–1062. [CrossRef]
77. Sanchez-Donoso, I.; Vilà, C.; Puigcerver, M.; Butkauskas, D.; de la Calle, J.R.C.; Morales-Rodríguez, P.A.; Rodríguez-Teijeiro, J.D. Are farm-reared quails for game restocking really common quails (*Coturnix coturnix*): A genetic approach. *PLoS ONE* **2012**, *7*, e39031. [CrossRef] [PubMed]
78. Oyun, N.Y.; Moiseyeva, I.G.; Sevastianova, A.A.; Vakhrameev, A.B.; Alexandrov, A.V.; Kuzevanova, A.Y.; Alimov, A.A.; Sulimova, G.E. Mitochondrial DNA polymorphism in different populations of Spangled Orloff chickens. *Genetika* **2015**, *51*, 1057–1065. (In Russian with English Summary). [CrossRef]
79. Oyun, N.Y.; Moiseyeva, I.G.; Sevastianova, A.A.; Vakhrameev, A.B.; Alexandrov, A.V.; Kuzevanova, A.Y.; Alimov, A.A.; Sulimova, G.E. Mitochondrial DNA polymorphism in different populations of Orloff Spangled chicken breed. *Russ. J. Genet.* **2015**, *51*, 908–915. [CrossRef]
80. Kowalczyk, M.; Staniszewski, A.; Kamińska, K.; Domaradzki, P.; Horecka, B. Advantages, possibilities, and limitations of mitochondrial DNA analysis in molecular identification. *Folia Biol.* **2021**, *69*, 101–111. [CrossRef]
81. Romanov, M.N.; Weigend, S. Genetic Diversity in Chicken Populations Based on Microsatellite Markers. In Proceedings of the Conference "From Jay Lush to Genomics: Visions for Animal Breeding and Genetics", Ames, IA, USA, 16–18 May 1999; Dekkers, J.C.M., Lamont, S.J., Rothschild, M.F., Eds.; Iowa State University, Department of Animal Science: Ames, IA, USA, 1999; p. 174. Available online: https://web.archive.org/web/20050314091227/http://www.agbiotechnet.com/proceedings/jaylush.asp#34 (accessed on 25 September 2023).
82. Mohammadabadi, M.R.; Nikbakhti, M.; Mirzaee, H.R.; Shandi, M.A.; Saghi, D.A.; Romanov, M.N.; Moiseyeva, I.G. Genetic variability in three native Iranian chicken populations of the Khorasan province based on microsatellite markers. *Genetika* **2010**, *46*, 572–576. Available online: https://www.researchgate.net/publication/44661596_Genetic_variability_in_three_native_Iranian_chicken_populations_of_the_Khorasan_province_based_on_microsatellite_markers (accessed on 25 September 2023). [CrossRef]
83. Weigend, S.; Romanov, M.N.; Ben-Ari, G.; Hillel, J. Overview on the Use of Molecular Markers to Characterize Genetic Diversity in Chickens. In Proceedings of the XXII World's Poultry Congress & Exhibition, Participant List & Full Text CD + Book of Abstracts, Istanbul, Turkey, 8–13 June 2004; WPSA—Turkish Branch: Istanbul, Turkey, 2004; p. 192. Available online: https://www.researchgate.net/publication/372751440_Overview_on_the_use_of_molecular_markers_to_characterize_genetic_diversity_in_chickens (accessed on 25 September 2023).
84. Dementieva, N.V.; Shcherbakov, Y.S.; Tyshchenko, V.I.; Terletsky, V.P.; Vakhrameev, A.B.; Nikolaeva, O.A.; Ryabova, A.E.; Azovtseva, A.I.; Mitrofanova, O.V.; Peglivanyan, G.K.; et al. Comparative analysis of molecular RFLP and SNP markers in assessing and understanding the genetic diversity of various chicken breeds. *Genes* **2022**, *13*, 1876. [CrossRef] [PubMed]
85. Hou, H.; Wang, X.; Zhang, C.; Tu, Y.; Lv, W.; Cai, X.; Xu, Z.; Yao, J.; Yang, C. Genomic analysis of GBS data reveals genes associated with facial pigmentation in Xinyang blue-shelled layers. *Arch. Anim. Breed.* **2020**, *63*, 483–491. [CrossRef]
86. Zhu, F.; Cui, Q.Q.; Hou, Z.C. SNP discovery and genotyping using Genotyping-by-Sequencing in Pekin ducks. *Sci. Rep.* **2016**, *6*, 36223. [CrossRef]
87. Sun, Y.; Wu, Q.; Lin, R.; Chen, H.; Zhang, M.; Jiang, B.; Wang, Y.; Xue, P.; Gan, Q.; Shen, Y.; et al. Genome-wide association study for the primary feather color trait in a native Chinese duck. *Front. Genet.* **2023**, *14*, 1065033. [CrossRef] [PubMed]
88. Grzegorczyk, J.; Gurgul, A.; Oczkowicz, M.; Szmatoła, T.; Fornal, A.; Bugno-Poniewierska, M. Single nucleotide polymorphism discovery and genetic differentiation analysis of geese bred in Poland, using genotyping-by-sequencing (GBS). *Genes* **2021**, *12*, 1074. [CrossRef]
89. Deng, M.T.; Zhu, F.; Yang, Y.Z.; Yang, F.X.; Hao, J.P.; Chen, S.R.; Hou, Z.C. Genome-wide association study reveals novel loci associated with body size and carcass yields in Pekin ducks. *BMC Genom.* **2019**, *20*, 1. [CrossRef] [PubMed]
90. Mathur, S.; DeWoody, J.A. Genetic load has potential in large populations but is realized in small inbred populations. *Evol. Appl.* **2021**, *14*, 1540–1557. [CrossRef]
91. Lake, J.A.; Dekkers, J.C.; Abasht, B. Genetic basis and identification of candidate genes for wooden breast and white striping in commercial broiler chickens. *Sci. Rep.* **2021**, *11*, 6785. [CrossRef] [PubMed]

92. Moiseeva, I. Chicken genetic resources in Russia. *Ptitsevodstvo* **1995**, *5*, 12–15. (In Russian)
93. Moiseyeva, I.G. The state of poultry genetic resources in Russia. *Anim. Genet. Resour.* **1996**, *17*, 73–86. [CrossRef]
94. Weigend, S.; Romanov, M.N.; Rath, D. Methodologies to Identify, Evaluate and Conserve Poultry Genetic Resources. In Proceedings of the XXII World's Poultry Congress & Exhibition, Participant List & Full Text CD + Book of Abstracts, Istanbul, Turkey, 8–13 June 2004; WPSA—Turkish Branch: Istanbul, Turkey, 2004; p. 84. Available online: https://www.researchgate.net/publication/250917228_Methodologies_to_identify_evaluate_and_conserve_poultry_genetic_resources (accessed on 25 September 2023).
95. Sulimova, G.E.; Stolpovsky, Y.A.; Kashtanov, S.N.; Moiseeva, I.G.; Zakharov, I.A. Methods of managing the genetic resources of domesticated animals. In *Fundamentals of Biological Resource Management: Collection of Scientific Articles*; Rysin, L.P., Ed.; Partnership of Scientific Publications KMK LLC: Moscow, Russia, 2005; pp. 331–342. ISBN 5-87317-254-4. Available online: https://elibrary.ru/item.asp?id=50435256 (accessed on 25 September 2023). (In Russian)
96. Tagirov, M.T.; Tereshchenko, L.V.; Tereshchenko, A.V. Substantiation of the possibility of using primary germ cells as material for the preservation of poultry genetic resources. *Ptakhivnytstvo* **2006**, *58*, 464–473. Available online: https://www.researchgate.net/publication/342751269_Tagirov_MT_Teresenko_LV_Teresenko_AV_Obosnovanie_vozmoznosti_ispolzovania_pervicnyh_zarodysevyh_kletok_v_kacestve_materiala_dla_sohranenia_geneticeskih_resursov_ptic_Ptahivnictvo_Mizvid_temat_nauk_zb_ (accessed on 25 September 2023). (In Russian with English Summary).
97. German, N.Y.; Volkova, N.A.; Larionova, P.V.; Vetokh, A.N.; Volkova, L.A.; Sermyagin, A.A.; Shakhin, A.V.; Anshakov, D.V.; Fisinin, V.I.; Zinovieva, N.A. Genome-wide association studies of growth dynamics in quails *Coturnix coturnix*. *Sel'skokhozyaistvennaya Biol.* **2022**, *57*, 1136–1146. (In Russian with English Summary). [CrossRef]

animals

MDPI

Article

Wild Avian Gut Microbiome at a Small Spatial Scale: A Study from a Mediterranean Island Population of *Alectoris rufa*

Monica Guerrini, Dalia Tanini, Claudia Vannini * and Filippo Barbanera

Department of Biology, University of Pisa, Via A. Volta 4, 56126 Pisa, Italy; filippo.barbanera@unipi.it (F.B.)
* Correspondence: claudia.vannini@unipi.it; Tel.: +39-050-2211358

Simple Summary: Our study is one of the few comparative and within-a-species descriptions of microbiomes in wild non-passerine birds. Particularly, it focuses on red-legged partridges, which are medium-sized gamebirds inhabiting open dry countryside and low-intensity cultivations with a mix of fallow and uncultivated areas in southwestern Europe. We wanted to study microbes living in their gut as their occurrence and diversity may affect both survival and reproduction of these birds. We collected fresh red-legged partridge fecal pellets at different sites located on both the western (two) and eastern (one) sides of Elba Island (central Italy). Although most represented bacteria were the same in all the three sites, we found differences between western and eastern Elban subpopulations in terms of microbiome composition and diversity. This result might be related to locally diverging individual physiological needs and/or to different intensities in past releases of captive-bred birds between the two sides of Elba. Overall, we suggest that the two partridge subpopulations should be managed separately to avoid any loss or significant variation in their microbiome structure.

Abstract: This research is one of the few comparative descriptions at an intraspecific level of wild non-passerine microbiomes. We investigated for the first time the gut microbiome of red-legged partridges (*Alectoris rufa*) using fecal pellets in order to provide a more informed management. We focused on a small Italian population consisting of two demes (WEST, EAST) separated by about 20 km on the opposite sides of Elba Island. Given the small spatial scale, we set up a sampling protocol to minimize contamination from environmental bacteria, as well as differences due to variations in—among others—habitat, season, and age of feces, that could possibly affect the investigation of the three Elban sites. We found a significant divergence between the WEST and EAST Elban subpopulations in terms of microbial composition and alpha diversity. Although most represented bacterial phyla were the same in all the sites (*Firmicutes*, *Actinobacteria*, *Proteobacteria*, and *Bacteroidetes*), microbiomes displayed a much higher diversity in western than in eastern partridges. This result might be related to locally diverging individual physiological needs and/or to different intensities in past releases of captive-bred birds between the two sides of Elba. We suggest that the two subpopulations should be treated as distinct management units.

Keywords: host-microbiome associations; microbiome; non-invasive sampling; red-legged partridge; 16S rRNA metabarcoding

1. Introduction

The rapid development in cultivation-independent high-throughput sequencing techniques allowed the outbreak of ground-breaking research on what Woese [1] referred to as the 'sleeping giant' of biology: the microbial world. This process revealed the largely unforeseen role that microorganisms play in development, growth, and health of virtually all living beings since they harbor the so-called microbiome [2].

Among the vertebrates, it is well known that the microbiome of the gastrointestinal tract can affect survival and reproductive performance through its interactions not only with nutrition but also with both the physiology and immune system of the host [3–5]. Symbiotic microbes can indeed play a pivotal role in herbivores' digestion [6,7], in fulfilling specific nutritional requirements (e.g., in whales [8]), and in protection against pathogens (e.g., [9,10]). Perturbations or even disruption of microbial communities—most commonly of the gastrointestinal tract, the so-called dysbiosis (sensu [11])—are often associated with a health disorder and can potentially result in a significant decline of both survival rate and fitness of the host [12,13]. Despite its role in nutritional uptake, detoxification, immune function, and competitive exclusion of pathogens [14,15], the gut microbiome has been less explored in wild birds than in mammals [16], with the non-passerine species being studied very rarely [17,18]. Therefore, a better understanding of wild bird microbiomes necessarily implies widening the range of investigated taxa.

The red-legged partridge (*Alectoris rufa*, Phasianidae) is a medium-sized gamebird that inhabits open dry countryside and low-intensity cultivations with a mix of fallow and uncultivated areas. In mainland Europe, it occurs from the Iberian Peninsula across most of France to northwestern Italy. Three subspecies are recognized: *A. r. rufa*, native to France and Italy, and *A. r. hispanica* and *A. r. intercedens* from northwestern Spain and the remainder of the Iberian Peninsula, respectively. Nominated subspecies have also been historically introduced into several islands, for instance, Corsica (VI c., [19]), Great Britain (XVII c., [20]), and—to the easternmost edge of its range—Elba, in the Tuscan Archipelago (Figure 1a). This latter consists of seven main islands, with Elba—the third in Italy by size (223.5 km^2)—hosting a small, protected red-legged partridge population, which is the only natural (not from an ex situ facility), long-established (at least since the late 1700s but likely much earlier [21]), and self-sustaining (no supplementation since the mid-1990s, [22]) Italian resource of this species. Nonetheless, partridges have long been hunted on Elba and despite the establishment of the Tuscan Archipelago National Park (1996)—with the majority of the island territory set under strict protection—a demographic collapse eventually took place by the end of 1990s, when the occurrence of only 30–50 pairs was assessed (see [21] and references therein). Later, Chiatante and colleagues [23] could not estimate any reliable population density value due to the paucity of individuals.

In a previous study based on the use of microsatellite DNA markers [22], we assessed that western and eastern regions of the Island, which are separated by less than 20 km, host genetically diverging subpopulations. In this study, we collected samples from western (two) and eastern (one) sites of Elba Island to compare microbial communities associated with the local red-legged partridges. On the one hand, we wanted to provide one of the few comparative descriptions at an intraspecific level of wild non-passerine bird microbiomes. On the other hand, we sought to gain first microbial data that would be useful to improve the management of either wild or farmed *A. rufa* using the Elban population as a reference.

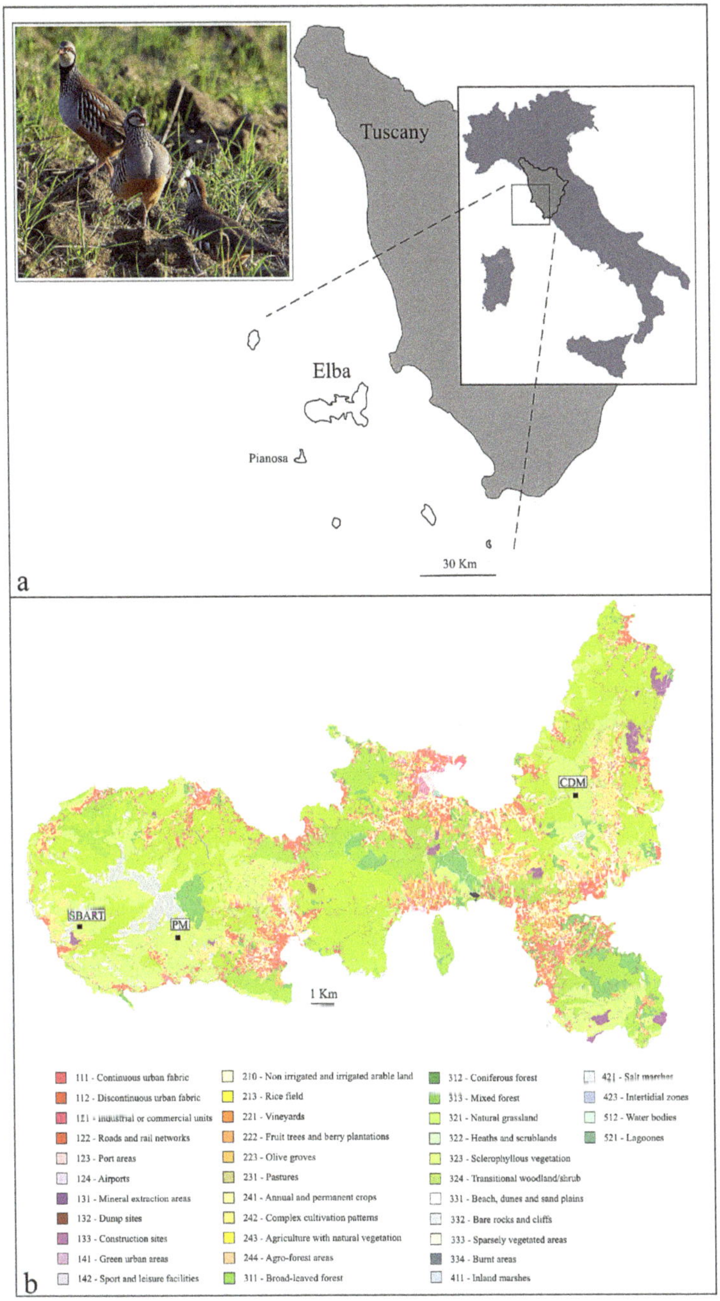

Figure 1. (**a**) The position of the island of Elba in Tuscany, central Italy; inset: a few red-legged partridges in the wild (courtesy: J.J. Negro, Spain). (**b**) Land cover map of Elba (source: Tuscan Region GEOscopio WMS: UCS10k 2019) obtained using QGIS v. 3.6 'Noosa'. The three sampling localities (same land class, n. 323) are indicated: SBART, San Bartolomeo; PM, Pietra Murata; CDM, Cima del Monte (see also Table 1).

Table 1. Sample information including name, locality, subpopulation, date, and elevation.

Sample	Locality	Subpopulation	Date	Elevation
SBART 13 SBART 14 SBART 19 SBART 20 SBART 21 SBART 22	San Bartolomeo	WEST	17 February 2019	402 m
PM 26 PM 41 PM 43 PM 44 PM 45 PM 46	Pietra Murata	WEST	8 February 2020	547 m
CDM 3A CDM 3B CDM 3C CDM 3MIX CDM 4A CDM 4B	Cima del Monte	EAST	15 December 2018	428 m

2. Materials and Methods

2.1. Biological Sampling

We selected three study sites within the limits of the National Park: two in the west (San Bartolomeo, 42°45′23.35″ lat. N and 10°07′31.56″ long. E, SBART; Pietra Murata, 42°45′15.03″ lat. N and 10°11′01.67″ long. E, PM) and one in the east (Cima del Monte, 42°47′51.74″ lat. N and 10°23′27.95″ long. E, CDM) Elba (Table 1). Located at a similar elevation (SBART, 402 m a.s.l.; PM, 547 m; CDM, 428 m) and holding the same exposure to the sun (open 360° view), these sites were assigned to the same land class (Figure 1b), which is dominated by herbaceous plants with garrigue and sclerophyllous vegetation among small grassland patches. They are largely characterized by *Brachypodium retusum* (Poaceae) and *Foeniculum vulgare* (Apiaceae), with *Phagnalon saxatile* (Asteraceae) and *Micromeria graeca* (Lamiaceae) occurring mainly in the CDM site. The lower herbaceous layer and the open spaces with meadows include *Poa annua* (Poaceae), *Trifolium subterraneum* (Fabaceae), *Hypochaeris achyrophorus* (Asteraceae), *Polycarpon tetraphyllum* (Caryophyllaceae), and *Plantago bellardii* (Plantaginaceae), whereas *Lavandula stoechas* (Lamiaceae) and *Cistus monspeliensis* (Cistaceae) are mainly found in the bushy areas, with *C. salviifolius* being more abundant on the western side of Elba [24,25]. Fresh (no more than 2–3 h old, to minimize contamination from environmental bacteria) and well-spaced fecal pellets were individually collected in winter (SBART, February 2019; PM, February 2020; CDM, December 2018) at least three days after the last rainfall, from 8.00 to 10.00 a.m., and with an air temperature ranging −2 to 4 °C. Separately kept in plastic vials (no chemicals added), samples were transported according to a strict cold chain until the final storage (−40 °C) at the University of Pisa. Six pellets for each site were randomly selected and used for the analyses (total sample size, 18).

2.2. DNA Extraction

Sample manipulations and DNA extractions were carried out under a dedicated sterile cabinet (Polaris 48, Steril Spa) and all materials and disposables were sterilized with UV light for 2 h. After the removal of a white layer (urine) that can be typically found on the top, genomic DNA was extracted using 200 mg of feces from a single pellet and the QIAamp® Fast DNA Stool Mini Kit (Qiagen, Hilden, Germany) according to the manufacturer's instructions with minor changes. We added 1 mL of InhibitEX® buffer, then we heated 5 min at 95 °C to improve the lysis, and the final elution was in 120 μL of Buffer ATE.

A blank extraction (no sample) was included in each session. DNA concentration was assessed with a Qubit 2.0 fluorometer and Qubit dsDNA HS Assay Kit (Thermo Fisher Scientific, Waltham, MA, USA).

2.3. 16S rRNA Gene Amplification and Sequencing

Amplifications were carried out under a different sterile cabinet (Top Safe 1.2, BioAir), with all materials, disposables, and the surface sterilized with UV light for 2 h before the setup of reactions. The V3–V4 regions of the 16S rRNA gene were amplified using the (5′-3′) primers (forward) 341F-CCTACGGGNGGCWGCAG and (reverse) 785R-GACTACHVGGGTATCTAATCC of [26]. The Illumina overhang adapter sequences added to the forward and reverse primers were TCGTCGGCAGCGTC AGATGTGTATAAGA-GACAG and GTCTCGTG GGCTCGGAGATGTGTA TAAGAGACAG, respectively. Each PCR was performed in 50 μL using the KAPA HiFi HotStart Ready Mix (Roche Diagnostics, Pleasanton, CA, USA), 8 ng of DNA, 0.2 μM of each primer, and adding 0.3 μg/μL of Ultrapure BSA (Bovine Serum Albumin, Invitrogen, Waltham, MA, USA). The thermal profile comprised 5 min at 95 °C, followed by 40 cycles of 30 s at 95 °C, 30 s at 55 °C, and 30 s at 72 °C, and a final extension at 72 °C for 5 min. The concentration of the amplicons was quantified using a Qubit fluorometer. Then, amplicons were barcoded and a sequencing library from each sample with an average concentration of 27.38 ng/μL was obtained. All 18 amplicons were sequenced on the Illumina MiSeq platform (sequencing depth of 2 × 100,000 paired-end reads) by IGATech (Udine, Italy).

2.4. Analysis of the Sequences

Raw reads of prokaryotic V3–V4 regions were analyzed using the QUANTITATIVE INSIGHTS INTO MICROBIAL ECOLOGY version 2 (QIIME2, https://qiime2.org, accessed on 1 March 2023) software package [27]. Reads were truncated at 260 bp length to remove the lower-quality last base calls. After that, quality filtering, primer trimming, pair-end read merging, and de novo chimera removal were performed with the divisive amplicon denoising algorithm (DADA2) plugin [28]. The resulting sequences were then used to generate amplicon sequence variants (ASVs). ASVs displaying a total abundance lower than 10 were discarded before proceeding with downstream analyses. ASV sequences were aligned with MAFFT [29] and a phylogenetic tree was inferred with FASTTREE [30]. This latter was manually inspected and no further chimeric sequence was disclosed. Taxonomic assignment of sequence variants was carried out using the release 132 of the Silva database [31]. A Naive Bayes classifier was trained extracting the regions of interest from SSU rRNA representative sequences (99% similarity clustered Operational Taxonomic Unit) as in [32]. Sequence variants identified as mitochondria, chloroplasts, unassigned, as well as all non-bacteria were removed before further data processing.

2.5. Statistical Analyses

All statistical analyses were performed using QIIME2 either for the three populations (SBART, PM, and CDM) and the western (WEST) and eastern (EAST) subpopulation. Taxa bar plots were produced at the phylum and genus level. Alpha and beta diversity were then estimated using ASVs. Rarefaction curves for each individual were obtained with a depth of 7701. Alpha diversity was assessed by calculating three different indexes: number of sequence variants (ASVs), Shannon's index (quantitative, non-phylogeny based index) for richness, and Pielou's Evenness for evenness. Comparison among index values for different communities was performed by the Kruskal–Wallis non-parametric test. Different metrics—Bray–Curtis and Jaccard for quantitative and qualitative data, respectively, and both weighted and unweighted Uni-Frac distances to assess the impact of phylogeny—were used for calculating beta diversity by means of a multivariate Principal Coordinates Analysis (PCoA) and Permanova (pairwise comparisons, 999 permutations).

3. Results

3.1. Sequencing Outcome

High-quality 16S rRNA gene sequences (2,092,950) were obtained from 18 fecal samples. The number of reads ranged 61,270–187,344 (116,275 ± 32,324, on average). After filtering all non-bacterial and unassigned sequences, the final dataset comprised 1,386,402 high-quality reads (77,022 ± 20,804, on average). The rarefaction curves reported in Figure S1 reached a plateau, thus confirming that the sequencing depth was sufficient to sample all variants in the libraries.

3.2. Composition of the Microbial Communities

The composition of the gut microbiome was reported for each partridge from the three study sites in the bar plots provided in Figure 2, in which differences between the WEST and EAST subpopulations concerning both the number of phyla and their relative abundance are visible. The western partridges (SBART and PM) showed a higher number of phyla and a more homogeneous distribution of the most represented ones (*Firmicutes, Actinobacteria, Proteobacteria*, and *Bacteroidetes*). On the other hand, *Firmicutes* was dominant in the EAST subpopulation with a relative abundance ranging 81–98% (Figure 2). The structure of the microbiome did not change when the relative abundance of taxa was investigated at the level of genus (Figure S2). The western partridges were characterized by a higher number and/or a more homogeneous distribution of taxa compared to the EAST subpopulation. In this, we found a lower number of genera along with a larger relative abundance of one (e.g., *Lactobacillus* in three out of six CDM samples) or of very few taxa than in the two western sites.

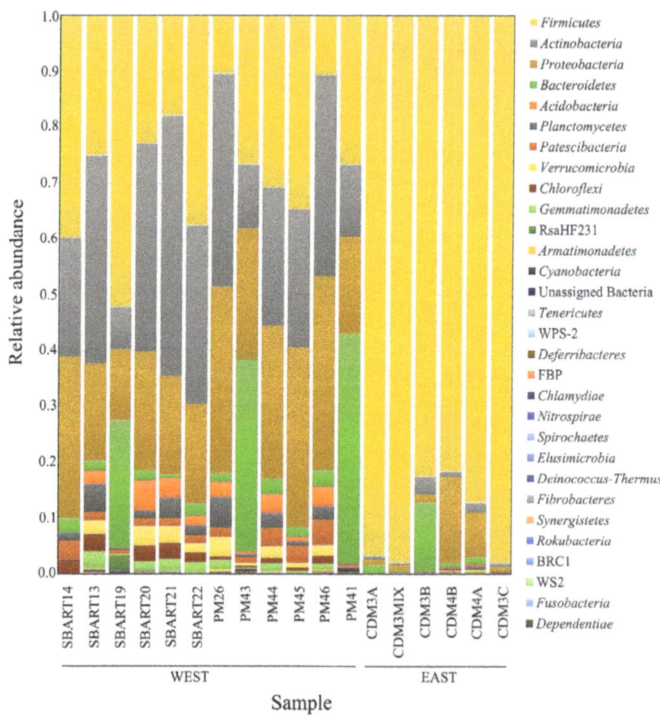

Figure 2. Relative abundances of bacterial phyla in the libraries as obtained for the three sampling sites. Each bar corresponds to one sample (single red-legged partridge).

3.3. Comparative Analyses of Microbial Diversity

Comparative estimates of alpha diversity (number of ASVs, Shannon, and Evenness indexes) for the three sites (SBART, PM, and CDM) and for the two subpopulations (WEST, EAST) were reported in Figure 3. The total number of ASVs was 3743, 3935, and 889 for SBART, PM, and CDM sites, respectively. The average number of ASVs was four times lower in CDM (148.17 ± 85.83) than in the other two localities (SBART, 623.83 ± 341.63; PM, 655.83 ± 216.47) (Figure 3). As far as the other two indexes are concerned, the average values obtained for SBART and PM were double than those of CDM (Shannon: SBART, 7.46 ± 1.10; PM, 7.67 ± 0.83; CDM, 3.03 ± 1.49; Pielou's Evenness: SBART 0.84 ± 0.06; PM, 0.83 ± 0.07; CDM, 0.42 ± 0.16; Figure 3). The Kruskal–Wallis tests (Table S1) returned highly significant differences ($p < 0.01$) for all alpha diversity indexes between each of the western sites (SBART and PM) and the eastern one (CDM) but not between SBART and PM within the WEST subpopulation ($p > 0.05$: Table S1). Likewise, the Kruskal–Wallis test indicated the occurrence of highly significant differences between WEST and EAST for all indexes ($p < 0.01$; Table S1).

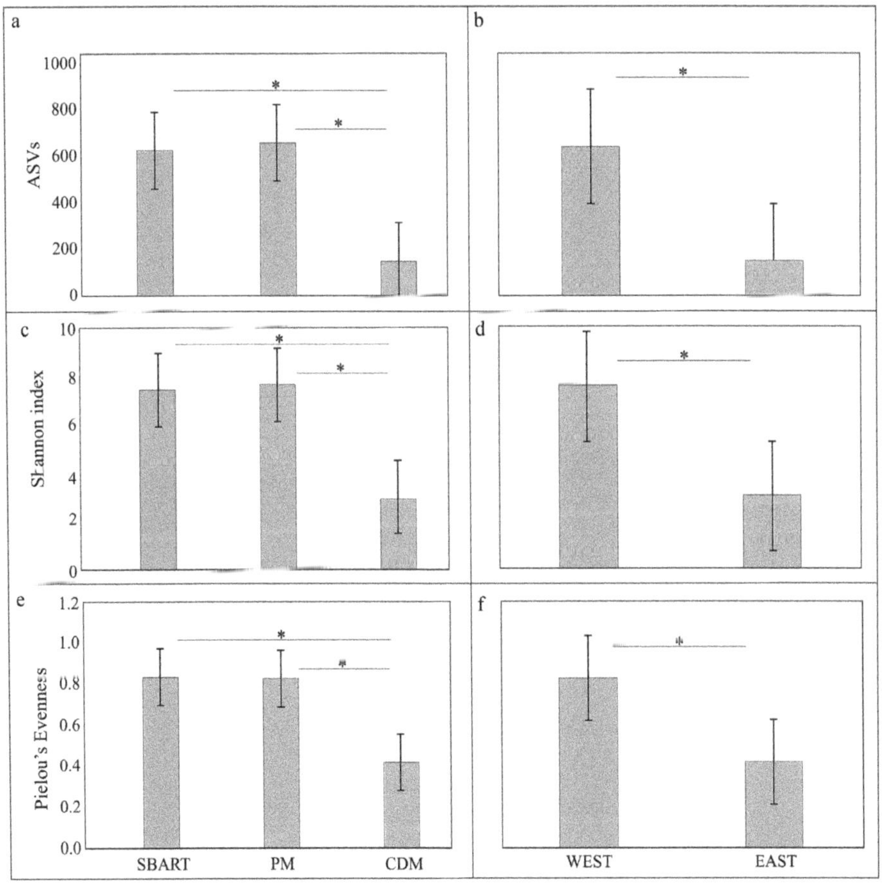

Figure 3. Measures of alpha diversity for the three sampling sites (SBART, PM, and CDM) and for each subpopulation (WEST and EAST) are reported as number of sequence variants, ASVs (**a**,**b**), Shannon (**c**,**d**), and Pielou's Evenness indexes (**e**,**f**). Error bars indicate the Standard Error (SE). Results of Kruskal–Wallis tests are reported in Table S1. Statistically significant comparisons ($p < 0.05$) are reported with an asterisk (*).

The PCoA of microbial communities as computed for all individuals using weighted Uni-Frac distances was reported in Figure 4 (similar results were obtained with Bray–Curtis, Jaccard, and unweighted Uni-Frac). Axes 1, 2, and 3 explained 84.13% of the total variability and a separation between the microbiomes from CDM and those from the two sites on the western side of Elba was disclosed. Permanova tests carried out for SBART, PM, CDM as well as for the two subpopulations (WEST, EAST) were reported in Table S2. All comparisons were statistically significant ($p < 0.01$) when Jaccard and Bray–Curtis distances were used—the two western sampling sites (SBART versus PM: $p = 0.011$) included—whereas with both weighted and unweighted Uni-Frac distances, SBART and PM were the only two sites that did not significantly diverge from each other ($p = 0.206$).

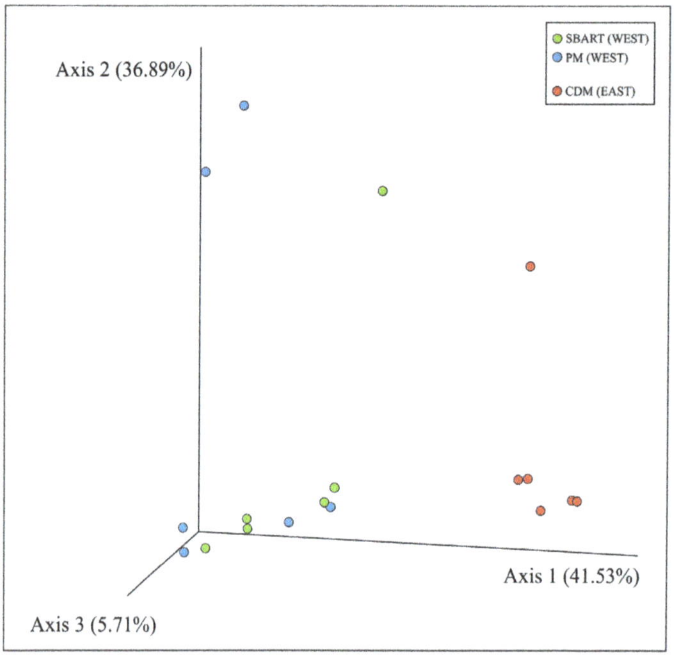

Figure 4. Principal Coordinates Analysis of microbial communities computed using weighted Uni-Frac distances and all available samples. Legend: green, SBART; blue, PM; red, CDM.

4. Discussion

Environmental conditions and diet are among the main factors shaping gut microbiome in wild passerine [33,34] and non-passerine [17,18] birds as well as in mammals. As to these latter, variation in the microbiome composition in wild populations, for instance, can be significantly driven by seasonal shifts in the diet [35–37]. Therefore, in the present work, consistency of habitat, foraging sources, and season among the sampling sites was deemed as a priority to mitigate the effects of exogenous factors. We investigated for the first time the gut microbiome of a non-model avian species: the red-legged partridge. According to the study of Turjeman and colleagues [38] in wild common cranes (*Grus grus*), non-invasive samples can even better represent host fecal microbial matter than those obtained from invasive ones as, for instance, there are no effects on the physiology of individuals due to the trapping. Hence, we used fecal pellets collected in localities within areas historically inhabited by *A. rufa* on the opposite sides of Elba Island [39] in herbaceous habitats—the preferred ones by partridges—that were highly consistent in terms of species' assembly (see Materials and Methods). Also, sites held a similar elevation, with this being another environmental parameter potentially influencing gut microbiome composition [40] (Table 1). Whereas partridges feed mainly—but not exclusively—on insects in the early stages of their life to meet protein requirements, they become increasingly herbivorous as soon as they become adults at 1 year of age [41]. The sampling was carried out in winter; hence, we most likely sampled adult partridges—after post-breeding dispersal—feeding mainly on fruits and seeds and secondarily on herbaceous plants (e.g., leathery leaves from evergreen species). Therefore, given the alleged age of host individuals and the high similarity in plant coverage (see Materials and Methods and Figure 1) of the sampled habitats, we made our very best to ensure the highest possible consistency in accordance with the aims of the work, which did not include the analysis of the diet of birds (see also the last part of Discussion). Considering another aspect, the fecal microbiome of birds is also known to be contributed to by different segments of the gastrointestinal tract. For instance, the cecum turned out to be one of the most important sources in the Japanese quail (*Coturnix japonica*, [42]). Although the domestic chicken (*Gallus gallus*) can expel the cecum content two or three times per day, there is little temporal continuity in the fecal microbiome growth and very poor is known about the involvement of other gastrointestinal regions in this as well as in other avian species [43]. Furthermore, the fecal microbial community is also dynamic over time once released in the field [44]. Therefore, we selected a strict time frame (8.00–10.00 a.m.) for the sampling at each site, so as to standardize any possible influence of different gastrointestinal tracts, and we collected only fresh pellets, namely those defecated no more than 2–3 h before—according to the experience of one of us (F.B.)—to minimize any possible contamination. Overall, this protocol represents a promising avenue to set up future similar investigations as comparative studies are important to provide knowledge of mechanisms affecting host–microbe relationships in the wild [45].

We reported microbiome composition and spatial structure in a wild *A. rufa* island population. We identified either conserved bacterial phyla or differences between subpopulations in terms of the microbial community across the Elban territory. The most common phyla were *Firmicutes* (53.8%), *Actinobacteria* (16.3%), *Proteobacteria* (15.1%), and *Bacteroidetes* (8.1%), which is a result in agreement with [16,33], where these four taxa were referred to as the 'core microbiome' of wild avian species, as well as with [46], where non-passerine captive birds were investigated. Within Phasianidae, *Firmicutes*, *Bacteroidetes*, and *Proteobacteria*—with *Tenericutes* as the fourth taxon in order of abundance—were the most represented phyla occurring in the microbial community of farmed Japanese quails [42]. Likewise, *Actinobacteria*, *Firmicutes*, *Proteobacteria*, and *Synergistetes* were the main microbial groups identified in wild ptarmigans (*Lagopus muta*, [18]). Overall, the composition of the Elban *A. rufa* microbiome, having *Firmicutes*, *Actinobacteria*, *Proteobacteria*, and *Bacteroidetes* as dominant phyla, was strongly consistent with the previous findings.

Firmicutes are Gram-positive bacteria producing short-chain fatty acids (butyrate) as byproducts of fermentation that can be directly absorbed by the host as a source of energy. Interestingly, they were very abundant in the CDM subpopulation, with *Lactobacillaceae* and *Erysipelotrichaceae* as the most widespread families. On the one hand, *Lactobacillus salivarius* can be used as a probiotic in the diet of captive-bred birds to improve nutrient absorption and increase growth performance (in domestic chickens and ducks, [47–49]). On the other hand, *Erysipelotrichaceae* are plentiful in domestic chickens converting feed to mass in an efficient manner [50]. Although it may be plausible that divergent physiological needs can account for the difference in the abundance of *Firmicutes* between EAST and WEST subpopulations, richness in lactobacilli might be (also) related to massive restocking with captive-bred individuals carried out on Elba especially between the 1960s and mid-1990s, with a higher intensity on the eastern than on the western side of the island [21,39]. This consideration might also explain the lower degree of microbiome diversity observed in the eastern subpopulation, as lots of evidence indicates that variability is usually lower in ex situ than in situ facilities. Nevertheless, this pattern is not straightforward as there are numerous study cases where the occurrence of no significant difference between captive and wild individuals was recorded as well, with even a few ones—focusing also on mammals and birds—where captive populations beneficiated instead from a higher microbiome diversity than wild ones [51]. Although the microbiome of farmed individuals usually changes in a short time after their release in the wild (e.g., [52–54]), we cannot exclude that some adaptation in the gut microbiome of EAST partridges may have been selected in captive-bred birds of various geographical origin—also imported from abroad [39]—during intense restocking. It is worth noting, indeed, that the eastern side of Elba is characterized by a dry Thermo-Mediterranean bioclimate, whereas a superior Meso-Mediterranean—shifting to humid Supra-Mediterranean—occurs on the western side, where less garrigue and rocky outcrops and more wooded patches can be comparatively found [22]. Unfortunately, birds employed for supplementation on Elba were obtained from farms that no longer exist today [39]; hence, there is no chance to carry out any ad hoc comparative study of captive versus wild bird microbiomes.

Despite the small spatial scale, a significant divergence was found between western (SBART and PM) and eastern (CDM) Elban *A. rufa* subpopulations, as shown by taxonomic assemblages as well as both alpha and beta diversity analyses. Partridges from SBART and PM sites displayed a higher number of taxa and a more homogeneous distribution of their relative dominances compared to those from CDM. Indeed, values of all alpha diversity indexes in the WEST sites were significantly higher than those recorded in the EAST subpopulation (Figures 2 and 3; Table S1). On the contrary, microbiomes of partridges sampled in SBART and PM sites accounted for a large homogeneity within the WEST subpopulation. This was also certified by Permanova tests and PCoA (Figure 4 and Table S2), thus pointing to the occurrence of distinct local adaptations by the two Elban *A. rufa* subpopulations.

The gut microbial communities turned out to be more closely related to each subpopulation (WEST, EAST: *c.* 20 km away one another) than to the sampling sites (SBART and PM within WEST). Interestingly, the same overall pattern was inferred by genotyping *A. rufa* partridges at a panel of 11 loci of the microsatellite DNA [22]. In this latter study, we investigated the partition of the genetic diversity among and within western and eastern Elban subpopulations and the results pointed to the occurrence of a well-established *A. rufa* spatial structure across the island. We found that the Elban population actually consisted of two demographically largely independent and genetically diverging groups resident in the opposite sides of the island. Overall, nuclear DNA diversity was high in the whole population—see also the lack of runs of homozygosity as inferred by [55]—and no significant difference in terms of allelic richness, Index of Nei, and observed heterozygosis was detected between the two subpopulations [22]. However, conventional wisdom suggests that the more the variables are controlled, the higher accuracy and precision of the survey will result. In the present study, we tried our very best to investigate the microbiomes of wild partridges in comparable conditions across the different sampling sites. However, an analysis of the diet of the sampled individuals (e.g., by mitochondrial metabarcoding of fecal pellets) was beyond the purposes of this study; hence, we are aware that we cannot assume that environmental factors were fully equivalent. Likewise, we cannot exclude that individual physiology (differences in the level of hormones, disease dynamics, level of carotenoids, etc., e.g., [56]) and sex may have also played some important role. Although 39 out of the 90 Elban samples investigated by Tanini and colleagues [22] were collected in the same three localities of the present study, regrettably, genetic and microbiome data were not obtained for the same individuals in the two studies. Altogether, these reasons can explain why the gut microbial communities of the two Elban subpopulations did not mirror host genetics (e.g., nuclear diversity estimates) for what concerns similarity in alpha diversity levels, even if they displayed very different community structures, thus reflecting the genetic divergence of the two *A. rufa* island demes.

Among many extrinsic and intrinsic factors [16], environment and host genetics are the main drivers shaping the gut microbiome composition in wild populations, with either of the two being prevalent according to different situations and organisms [40,45,57,58]. Several studies investigated the role played by host genetics, which appears to be stronger in mammals than in birds [59,60]. Nevertheless, most research focused on differences among distinct bird species [61,62], while microbiome shaping at a within-species level received much less attention so far. For example, Fleischer and collaborators [61] found little support for a large-scale control of the microbiome by host genetics, suggesting this would not necessarily imply the same at lower scales. In the present study, the correspondence we found between the spatial genetic and microbiome structure in the Elban partridges seems to point toward this direction, although further, specifically designed studies are needed to better investigate this issue.

The first investigation of the gut microbiome in wild *A. rufa* lays the foundations to improve some aspects in the management of this species. The Elban population—the only natural, long-established, and self-sustaining [22] of Italy—can work, indeed, as a reference in studies aimed at exploring the gut microbiome of *A. rufa*. A comparison with the most important farmed populations of Italy would be valuable for ex situ managers, for instance, to better understand the effect of the preventive supply of antimicrobials and coccidiostatics to captive birds until their release in nature [63]. Likewise, a comparison between the gut microbiome of native and introduced *A. rufa* should be pursued as well as the occurrence of red-legged partridges imported from abroad is a matter of fact across the species' range [64]. In particular, we agree with Lavretsky and co-authors [65], who successfully reported over a heavily managed and worldwide translocated gamebird, the mallard (*Anas platyrhynchos*), that reciprocal interactions between natural and captive-bred individuals—we add, also those occurring at the microbial scale—will ever increasingly lead to the rise of admixed populations in light of ongoing rapid habitat transformation and climate changes. Indeed, as in the mallard, introduced *A. rufa* populations are capable to adapt locally even within a few decades since farm releases (e.g., [66]).

5. Conclusions

We investigated for the first time the microbiome of *A. rufa*, focusing on a wild protected island population, and we found that both its composition and genetic diversity varied at a small spatial scale. The divergence between the microbes associated with birds living on the opposite sides of Elba Island agreed with the known spatial genetic structure of the same *A. rufa* population. As we already suggested in this previous study, we advise the National Park that the two subpopulations should be managed separately also to avoid any loss and/or significant homogenization in the microbiome structure of the Elban population.

Supplementary Materials: The following supporting information can be downloaded at: https://www.mdpi.com/article/10.3390/ani13213341/s1, Table S1. Kruskal Wallis tests for alpha diversity indexes among the three sampling sites (SBART, PM, and CDM) and the two subpopulations (WEST, EAST). Statistically significant comparisons are reported in bold ($p < 0.05$, see also Figure 3). Table S2. Permanova tests among the three sampling sites (SBART, PM, and CDM) and the two subpopulations (WEST, EAST) carried out with different metrics (999 permutations). Statistically significant comparisons are reported in bold ($p < 0.05$). Figure S1. Rarefaction curves of alpha diversity approaching the saturation plateau (sampling depth, 7701). Each sample with relative color is reported in the box to the right side. Presence and abundance of the genus *Faecalicatea* is indicated by an asterisk. Figure S2. Relative abundances of bacterial genera in the libraries as obtained for the three sampling sites. Each bar corresponds to one sample (single red-legged partridge).

Author Contributions: C.V. and F.B. conceived and designed the study. Sampling was carried out by F.B. Data collection and analysis were performed by M.G., C.V., and D.T. The first draft of the manuscript was written by M.G. and F.B. and all authors commented on previous versions of the manuscript. All authors have read and agreed to the published version of the manuscript.

Funding: This work was supported by Fondazione Isola d'Elba Onlus (Portoferraio, Livorno) with a research grant to F.B. (16 September 2020).

Institutional Review Board Statement: This research relied on non-invasively collected bird fecal pellets and, as such, procedures employed did not require approval from the Animal Welfare Body ('Organismo preposto al Benessere Animale', in Italian) of the University of Pisa.

Informed Consent Statement: Not applicable.

Data Availability Statement: Sequences (raw reads) were deposited in the European Nucleotide Archive (ENA) at EMBL-EBI under study accession number PRJEB59418.

Acknowledgments: The authors are also grateful to Fabrizio Erra and Francesco Paolo Frontini (Dipartimento di Biologia, Università di Pisa) for their collaboration in preparing line drawings and laboratory work, respectively. The authors also thank Angelino Carta (Dipartimento di Biologia, Università di Pisa) for the productive discussions on the vegetation and land use of Elba Island.

Conflicts of Interest: The authors have no relevant financial or non-financial interest to disclose.

References

1. Woese, C.R. Default taxonomy: Ernst Mayr's view of the microbial world. *Proc. Natl. Acad. Sci. USA* **1998**, *95*, 1043–11046. [CrossRef]
2. Shropshire, J.D.; Bordenstein, S.R. Speciation by symbiosis: The microbiome and behavior. *mBio* **2016**, *7*, e01785-15. [CrossRef]
3. Lee, W.J.; Hase, K. Gut microbiota-generated metabolites in animal health and disease. *Nat. Chem. Biol.* **2014**, *10*, 416–424. [CrossRef]
4. Amato, K.R.; Metcalf, J.L.; Song, S.J.; Hale, V.L.; Clayton, J.; Ackermann, G.; Humphrey, G.; Niu, K.; Cui, D.; Zhao, H.; et al. Using the gut microbiota as a novel tool for examining colobine primate GI health. *Glob. Ecol. Conserv.* **2016**, *7*, 225–237. [CrossRef]
5. Petrelli, S.; Buglione, M.; Rivieccio, E.; Ricca, E.; Baccigalupi, L.; Scala, G.; Fulgione, D. Reprogramming of the gut microbiota following feralization in Sus scrofa. *Anim. Microbiome* **2023**, *5*, 14. [CrossRef]
6. Dehority, B.A. *Rumen Microbiology*; Nottingham University Press: Nottingham, UK, 2003.
7. Kartzinel, T.R.; Hsing, J.C.; Musili, P.M.; Brown, B.R.P.; Pringle, R.M. Covariation of diet and gut microbiome in African megafauna. *Proc. Natl. Acad. Sci. USA* **2019**, *116*, 23588–23593. [CrossRef]
8. Olsen, M.A.; Blix, A.S.; Utsi, T.H.A.; Sørmo, W.; Mathiesen, S.D. Chitinolytic bacteria in the minke whale forestomach. *Can. J. Microbiol.* **2000**, *46*, 85–94. [CrossRef]
9. Silva, A.M.; Barbosa, F.H.; Duarte, R.; Vieira, L.Q.; Arantes, R.M.; Nicoli, J.R. Effect of *Bifidobacterium longum* ingestion on experimental salmonellosis in mice. *J. Appl. Microbiol.* **2004**, *97*, 29–37. [CrossRef]
10. Ley, R.; Turnbaugh, P.; Klein, S.; Gordon, J.I. Human gut microbes associated with obesity. *Nature* **2006**, *444*, 1022–1023. [CrossRef]
11. West, A.G.; Waite, D.W.; Deines, P.; Bourne, D.G.; Digby, A.; McKenzie, V.J.; Taylor, M.W. The microbiome in threatened species conservation. *Biol. Conserv.* **2019**, *229*, 85–98. [CrossRef]
12. Gilbert, J.A.; Quinn, R.A.; Debelius, J.; Xu, Z.Z.; Morton, J.; Garg, N.; Kansson, J.K.; Dorrestein, P.C.; Knight, R. Microbiome-wide association studies link dynamic microbial consortia to disease. *Nature* **2016**, *535*, 94–103. [CrossRef] [PubMed]
13. McKenzie, V.J.; Kueneman, J.G.; Harris, R.N. Probiotics as a tool for disease mitigation in wildlife: Insights from food production and medicine. *Ann. N. Y. Acad. Sci.* **2018**, *1429*, 18–30. [CrossRef]
14. Kimura, N.; Yoshikane, M.; Kobayashi, A. Microflora of the bursa of Fabricius of chickens. *Poult. Sci.* **1986**, *65*, 1801–1807. [CrossRef] [PubMed]
15. Godoy-Vitorino, F.; Ley, R.E.; Gao, Z.; Pei, Z.; Ortiz-Zuazaga, H.; Pericchi, L.R.; Garcia-Amado, M.A.; Michelangeli, F.; Blaser, M.J.; Gordon, J.I.; et al. Bacterial community in the crop of the hoatzin, a neotropical folivorous flying bird. *Appl. Environ. Microbiol.* **2008**, *74*, 5905–5912. [CrossRef]
16. Grond, K.; Sandercock, B.K.; Jumpponen, A.; Zeglin, L.H. The avian gut microbiota: Community, physiology and function in wild birds. *J. Avian Biol.* **2018**, *49*, e01788. [CrossRef]
17. Roggenbuck, M.; Bærholm Schnell, I.; Blom, N.; Bælum, J.; Bertelsen, M.F.; Sicheritz-Pontén, T.; Sørensen, S.J.; Gilbert, M.T.P.; Graves, G.R.; Hansen, L.H. The microbiome of New World vultures. *Nat. Commun.* **2014**, *5*, 5498. [CrossRef] [PubMed]
18. Salgado-Flores, A.; Tveit, A.T.; Wright, A.-D.; Pope, P.B.; Sundset, M.A. Characterization of the cecum microbiome from wild and captive rock ptarmigans indigenous to Arctic Norway. *PLoS ONE* **2019**, *14*, e0213503. [CrossRef] [PubMed]
19. Barbanera, F.; Forcina, G.; Guerrini, M.; Dini, F. Molecular phylogeny and diversity of Corsican red-legged partridge: Hybridization and management issues. *J. Zool.* **2011**, *285*, 56–65. [CrossRef]
20. Barbanera, F.; Forcina, G.; Cappello, A.; Guerrini, M.; van Grouw, H.; Aebischer, N.J. Introductions over introductions: The genomic adulteration of an early genetically valuable alien species in the United Kingdom. *Biol. Invasions* **2015**, *17*, 409–422. [CrossRef]
21. Barbanera, F. On the origins and history of the red-legged partridge (*Alectoris rufa*) from Elba Island (Tuscan Archipelago, Italy). *Atti Soc. Toscana Sci. Nat. Mem. Serie B* **2021**, *128*, 45–55. [CrossRef]
22. Tanini, D.; Guerrini, M.; Vannini, C.; Barbanera, F. Unexpected genetic integrity boosts hope for the conservation of the red-legged partridge (*Alectoris rufa*, Galliformes) in Italy. *Zoology* **2022**, *155*, 126056. [CrossRef] [PubMed]

23. Chiatante, G.; Meriggi, A.; Giustini, D.; Baldaccini, N.E. Density and habitat requirements of red-legged partridge on Elba Island (Tuscan Archipelago, Italy). *Ital. J. Zool.* **2013**, *80*, 402–411. [CrossRef]
24. Foggi, B.; Cartei, L.; Pignotti, L.; Signorini, M.A.; Viciani, D. Il paesaggio vegetale dell'Isola d'Elba (Arcipelago toscano): Studio fitosociologico e cartografico. *Fitosociologia* **2006**, *43*, 3–94. (In Italian)
25. Carta, A.; Taboada, T.; Müller, J.V. Diachronic analysis using aerial photographs across fifty years reveals significant land use and vegetation changes on a Mediterranean island. *Appl. Geogr.* **2018**, *98*, 78–86. [CrossRef]
26. Klindworth, A.; Pruesse, E.; Schweer, T.; Peplies, J.; Quast, C.; Horn, M.; Glöckner, F.O. Evaluation of general 16S ribosomal RNA gene PCR primers for classical and next-generation sequencing-based diversity studies. *Nucleic Acids Res.* **2013**, *41*, e1. [CrossRef]
27. Bolyen, E.; Rideout, J.R.; Dillon, M.R.; Bokulich, N.A.; Abnet, C.C.; Al-Ghalith, G.A.; Alexander, H.; Alm, E.J.; Arumugam, M.; Asnicar, F.; et al. Reproducible, interactive, scalable and extensible microbiome data science using QIIME. *Nat. Biotechnol.* **2019**, *37*, 852–857. [CrossRef]
28. Callahan, B.J.; McMurdie, P.J.; Rosen, M.J.; Han, A.W.; Johnson, A.J.A.; Holmes, S.P. DADA2: High-resolution sample inference from Illumina amplicon data. *Nat. Methods* **2016**, *13*, 581–583. [CrossRef]
29. Katoh, K.; Standley, D.M. MAFFT multiple sequence alignment software version 7: Improvements in performance and usability. *Mol. Biol. Evol.* **2013**, *30*, 72–780. [CrossRef]
30. Price, M.N.; Dehal, P.S.; Arkin, A.P. FastTree 2—Approximately maximum-likelihood trees for large alignments. *PLoS ONE* **2010**, *5*, e9490. [CrossRef]
31. Quast, C.; Pruesse, E.; Yilmaz, P.; Gerken, J.; Schweer, T.; Yarza, P.; Peplies, J.; Glöckner, F.O. The SILVA ribosomal RNA gene database project: Improved data processing and web-based tools. *Nucleic Acids Res.* **2013**, *41*, D590–D596. [CrossRef]
32. Werner, J.J.; Koren, O.; Hugenholtz, P.; DeSantis, T.Z.; Walters, W.A.; Caporaso, J.G.; Angenent, L.T.; Knight, R.; Ley, R.E. Impact of training sets on classification of high-throughput bacterial 16S rRNA gene surveys. *ISME J.* **2012**, *6*, 94–103. [CrossRef]
33. Hird, S.M.; Sánchez, C.; Carstens, B.C.; Brumfield, R.T. Comparative Gut microbiota of 59 Neotropical Bird Species. *Front. Microbiol.* **2015**, *6*, 1403. [CrossRef]
34. Schmiedová, L.; Tomášek, O.; Pinkasová, H.; Albrecht, T.; Kreisinger, J. Variation in diet composition and its relation to gut microbiota in a passerine bird. *Sci. Rep.* **2022**, *12*, 3787. [CrossRef] [PubMed]
35. Carey, H.V.; Walters, W.A.; Knight, R. Seasonal restructuring of the ground squirrel gut microbiota over the annual hibernation cycle. *Am. J. Physiol. Regul. Integr. Comp. Physiol.* **2013**, *304*, R33–R42. [CrossRef] [PubMed]
36. Maurice, C.F.; Knowles, S.C.; Ladau, J.; Pollard, K.S.; Fenton, A.; Pedersen, A.B.; Turnbaugh, P.J. Marked seasonal variation in the wild mouse gut microbiota. *ISME J.* **2015**, *9*, 2423–2434. [CrossRef] [PubMed]
37. Xue, Z.; Zhang, W.; Wang, L.; Hou, R.; Zhang, M.; Fei, L.; Zhang, X.; Huang, H.; Bridgewater, L.; Jiang, Y.; et al. The bamboo-eating giant panda harbors a carnivore-like gut microbiota, with excessive seasonal variations. *mBio* **2015**, *6*, e00022-15. [CrossRef] [PubMed]
38. Turjeman, S.; Pekarsky, S.; Corl, A.; Kamath, P.L.; Getz, W.M.; Bowie, R.C.K.; Markin, Y.; Nathan, R. Comparing invasive and noninvasive faecal sampling in wildlife microbiome studies: A case study on wild common cranes. *Mol. Ecol. Res.* **2023**, *23*, 359–367. [CrossRef]
39. Forcina, G.; Guerrini, M.; Barbanera, F. Non-native and hybrid in a changing environment: Conservation perspectives for the last Italian red-legged partridge (*Alectoris rufa*) population with long natural history. *Zoology* **2020**, *138*, 125740. [CrossRef]
40. Suzuki, T.A.; Martins, F.M.; Nachman, M.W. Altitudinal variation of the gut microbiota in wild house mice. *Mol. Ecol.* **2018**, *9*, 2378–2390. [CrossRef]
41. Spanò, S. *La Pernice Rossa*; Edizioni Il Piviere: Gavi, Italy, 2010. (In Italian)
42. Wilkinson, N.; Hughes, R.J.; Aspden, W.J.; Chapman, J.; Moore, R.J.; Stanley, D. The gastrointestinal tract microbiota of the Japanese quail, *Coturnix japonica*. *Appl. Microbiol. Biotechnol.* **2016**, *100*, 4201–4209. [CrossRef]
43. Sekelja, M.; Rud, I.; Knutsen, S.H.; Denstadli, V.; Westereng, B.; Næs, T.; Rudi, K. Abrupt Temporal Fluctuations in the Chicken Fecal Microbiota Are Explained by Its Gastrointestinal Origin. *Appl. Environ. Microbiol.* **2012**, *78*, 2941–2948. [CrossRef] [PubMed]
44. Hale, V.L.; Tan, C.L.; Niu, K.; Yang, Y.; Cui, D.; Zhao, H.; Knight, R.; Amato, K.R. Effects of field conditions on fecal microbiota. *J. Microbiol. Methods* **2016**, *130*, 180–188. [CrossRef] [PubMed]
45. Steury, R.A.; Currey, M.C.; Cresko, W.A.; Bohannan, B.J.M. Population Genetic Divergence and Environment Influence the Gut Microbiome in Oregon Threespine Stickleback. *Genes* **2019**, *10*, 484. [CrossRef] [PubMed]
46. Xiao, K.; Fan, Y.; Zhang, Z.; Shen, X.; Li, X.; Liang, X.; Bi, R.; Wu, Y.; Zhai, J.; Dai, J.; et al. Covariation of the Fecal Microbiome with Diet in Nonpasserine Birds. *mSphere* **2021**, *6*, e00308-21. [CrossRef]
47. Angelakis, E.; Raoult, D. The increase of *Lactobacillus* species in the gut flora of newborn broiler chicks and ducks is associated with weight gain. *PLoS ONE* **2010**, *5*, e10463. [CrossRef]
48. Awad, W.A.; Ghareeb, K.; Böhm, J. Effect of addition of a probiotic micro-organism to broiler diet on intestinal mucosal architecture and electrophysiological parameters. *J. Anim. Physiol. Anim. Nutr.* **2010**, *94*, 486–494. [CrossRef]
49. Garcia-Mazcorro, J.F.; Castillo-Carranza, S.A.; Guard, B.; Gomez-Vazquez, J.P.; Dowd, S.E.; Brigthsmith, D.J. Comprehensive Molecular Characterization of Bacterial Communities in Feces of Pet Birds Using 16S Marker Sequencing. *Microb. Ecol.* **2017**, *73*, 224–235. [CrossRef]

50. Stanley, D.; Hughes, R.J.; Geier, M.S.; Moore, R.J. Bacteria within the Gastrointestinal Tract Microbiota Correlated with Improved Growth and Feed Conversion: Challenges Presented for the Identification of Performance Enhancing Probiotic Bacteria. *Front. Microbiol.* **2016**, *7*, 187. [CrossRef]
51. Hauffe, H.C.; Barelli, C. Conserve the germs: The gut microbiota and adaptive potential. *Conserv. Genet.* **2019**, *20*, 19–27. [CrossRef]
52. Chong, R.; Grueber, C.E.; Fox, S.; Wise, P.; Barrs, V.R.; Hogg, C.J.; Belov, K. Looking like the locals—Gut microbiome changes post-release in an endangered species. *Anim. Microbiome* **2019**, *1*, 8. [CrossRef]
53. Quiroga-González, C.; Cardenas, L.A.C.; Ramírez, M.; Reyes, A.; González, C.; Stevenson, P.R. Monitoring the variation in the gut microbiota of captive woolly monkeys related to changes in diet during a reintroduction process. *Sci. Rep.* **2021**, *11*, 6522. [CrossRef]
54. Eliades, S.J.; Brown, J.C.; Colston, T.J.; Fisher, R.N.; Niukula, J.B.; Gray, K.; Vadada, J.; Rasalato, S.; Siler, C.D. Gut microbial ecology of the Critically Endangered Fijian crested iguana (*Brachylophus vitiensis*): Effects of captivity status and host reintroduction on endogenous microbiomes. *Ecol. Evol.* **2021**, *11*, 4731–4743. [CrossRef]
55. Chattopadhyay, B.; Forcina, G.; Garg, K.M.; Irestedt, M.; Guerrini, M.; Barbanera, F.; Rheindt, F.E. Novel genome reveals susceptibility of popular gamebird, the red-legged partridge (*Alectoris rufa*, Phasianidae), to climate change. *Genomics* **2021**, *113*, 3430–3438. [CrossRef]
56. García, J.T.; Pérez-Rodríguez, L.; Calero-Riestra, M.; Sánchez-Barbudo, I.; Viñuela, J.; Casas, F. Sexual differences in blood parasite infections, circulating carotenoids and body condition in free-living red-legged partridges. *J. Zool.* **2023**, *320*, 260–270. [CrossRef]
57. Ren, T.; Boutin, S.; Humphries, M.M.; Dantzer, B.; Gorrel, J.C.; Coltman, D.W.; McAdam, A.G.; Wu, M. Seasonal, spatial, and maternal effects on gut microbiome in wild red squirrels. *Microbiome* **2017**, *5*, 163. [CrossRef]
58. Hicks, A.L.; Lee, K.J.; Couto-Rodriguez, M.; Patel, J.; Sinha, R.; Guo, C.; Olson, S.H.; Seimon, A.; Seimon, T.A.; Ondzie, A.U.; et al. Gut microbiomes of wild great apes fluctuate seasonally in response to diet. *Nat. Commun.* **2018**, *9*, 1786. [CrossRef]
59. Song, S.J.; Sanders, J.G.; Delsuc, F. Comparative Analyses of Vertebrate Gut Microbiomes Reveal Convergence between Birds and Bats. *mBio* **2020**, *11*, e02901-19. [CrossRef]
60. Trevelline, B.K.; Khol, K.D. The gut microbiome influences host diet selection behavior. *Proc. Natl. Acad. Sci. USA* **2020**, *119*, 2117537119. [CrossRef]
61. Fleischer, R.; Risely, A.; Hoeck, P.E.A.; Keller, L.F.; Sommer, S. Mechanisms governing avian phylosymbiosis: Genetic dissimilarity based on neutral and MHC regions exhibits little relationship with gut microbiome distributions of Galápagos mockingbirds. *Ecol. Evol.* **2020**, *10*, 13345–13354. [CrossRef]
62. Bodawatta, K.H.; Koane, B.; Maiah, G.; Sam, K.; Poulsen, M.; Jønsson, K.A. Species-specific but not phylosymbiotic gut microbiomes of New Guinean passerine birds are shaped by diet and flight-associated gut modifications. *Proc. R. Soc. B Biol. Sci.* **2021**, *288*, 20210446. [CrossRef]
63. Díaz-Sánchez, S.; Höfle, U.; Villanúa, D.; Gortázar, C. Health Monitoring and Disease Control in Red-Legged Partridges. In *The Future of the Red-Legged Partridge. Science, Hunting and Conservation*; Casas, F., García, J.T., Eds.; Springer Nature: Basel, Switzerland, 2022. [CrossRef]
64. Forcina, G.; Tang, Q.; Cros, E.; Guerrini, M.; Rheindt, F.E.; Barbanera, F. Genome wide markers redeem the identity of a heavily managed and doomed-too-early gamebird. *Proc. R. Soc. B Biol. Sci.* **2021**, *288*, 20210285. [CrossRef]
65. Lavretsky, P.; Mohl, J.E.; Söderquist, P.; Kraus, R.H.S.; Schummer, M.L.; Brown, J.I. The meaning of wild: Genetic and adaptive consequences from large-scale releases of domestic mallards. *Commun. Biol.* **2023**, *6*, 819. [CrossRef]
66. Baratti, M.; Ammannati, M.; Magnelli, C.; Dessì-Fulgheri, F. Introgression of chukar genes into a reintroduced red-legged partridge (*Alectoris rufa*) population in central Italy. *Anim. Genet.* **2004**, *36*, 29–35. [CrossRef]

animals

MDPI

Article

Developing Methods for Maintaining Genetic Diversity in Novel Aquaculture Species: The Case of *Seriola lalandi*

Víctor Martinez [1,*], Nicolas Galarce [2] and Alvin Setiawan [3]

[1] INBIOGEN, Department of Animal Production, Faculty of Veterinary Sciences, Universidad de Chile, Avda. Santa Rosa 11735, Santiago 8820808, Chile
[2] Escuela de Medicina Veterinaria, Facultad de Ciencias de la Vida, Universidad Andrés Bello, Santiago 8370146, Chile
[3] Northland Aquaculture Centre, National Institute of Water and Atmospheric Research, Ruakaka 0116, New Zealand
* Correspondence: vmartine@uchile.cl

Simple Summary: Limiting inbreeding rates in farmed populations is crucial to ensuring long-term commercial viability. This task is particularly challenging in the aquaculture of mass communal spawning species, such as the yellowtail kingfish (*Seriola lalandi*). This reproductive strategy often results in a skewed parental genetic contribution while introducing additional complexities in parentage determination (c.f., controlled matings). To overcome these issues, we developed a marker panel based on genotyping-by-sequencing spanning 300 SNPs for parentage determination. Panel performance was satisfactory, which advocates for its employment to increase the long-term sustainability of this aquaculture resource when implementing breeding programs.

Abstract: Developing sound breeding programs for aquaculture species may be challenging when matings cannot be controlled due to communal spawning. We developed a genotyping-by-sequencing marker panel of 300 SNPs for parentage testing and sex determination by using data from an in-house reference genome as well as a 90 K SNP genotyping array based on different populations of yellowtail kingfish (*Seriola lalandi*). The minimum and maximum distance between adjacent marker pairs were 0.7 Mb and 13 Mb, respectively, with an average marker spacing of 2 Mb. Weak evidence of the linkage disequilibrium between adjacent marker pairs was found. The results showed high panel performance for parental assignment, with probability exclusion values equaling 1. The rate of false positives when using cross-population data was null. A skewed distribution of genetic contributions by dominant females was observed, thus increasing the risk of higher rates of inbreeding in subsequent captive generations when no parentage data are used. All these results are discussed in the context of breeding program design, using this marker panel to increase the sustainability of this aquaculture resource.

Keywords: *seriola*; inbreeding; genotyping-by-sequencing

1. Introduction

The yellowtail kingfish (*Seriola lalandi*) is a pelagic carnivorous fish that inhabits tropical and temperate waters of the Southern Hemisphere and the Northern Pacific, with known populations in Australia, New Zealand, Japan [1], the Southeast China Sea [2], the Mediterranean Sea [3], and the Pacific coast of South America [4–6]. As the demand for yellowtail kingfish continuously grows and fishery quotas have reached maximum levels [3], commercial aquaculture production of this species has been successfully established in Australia [7], the Netherlands, and Denmark, while the establishment of new farms has been planned in New Zealand [8] as well as North and South America [9,10].

The complete production cycle of *S. lalandi* has been successfully set up along the northern coast of Chile [10]. However, relatively little attention has been paid to a proper understanding of the genetic structure and diversity of both natural and captive populations of this species. In countries such as Germany, the Netherlands, and the USA, commercial aquaculture has been established using juveniles from single farms (particularly from Chile, Juan Lacamara, Pers. Comm.), where broodstock were initially sourced from natural fisheries, often with little or no attention to genetic diversity [4]. Nevertheless, such practices are not compatible with the long-term viability of the industry [11].

Traditional pedigree-based breeding programs for *S. lalandi* can also prove challenging. Similar to many other species, *S. lalandi* is a communal broadcast spawner that readily breeds in captivity without requiring hormonal inductions or gamete stripping for in-vitro fertilization (IVF) [7]. While this trait is highly beneficial for practical husbandry and production purposes, the absence of controlled matings prevents the holding of progeny by family, which allows for simple (non-molecular method) pedigree tracking. Combined with their high fecundity rate and propensity for skewed parental contribution [7], there is a high risk that high inbreeding rates arise in breeding programs, leading to a rapid reduction in effective population size (Ne) [12–17]. Furthermore, the difficulty in securing an even representation of all the broodstock in single production batches adds complexity for accurately assessing genetic parameters. For example, heritability and genetic correlations for harvest traits estimated in a commercial population were in most cases not statistically significant, which is likely due to the low number of parents (8 sires and 6 dams) used for estimation of such parameters [18,19].

For this reason, it is necessary to develop accurate marker panels to assess parentage while using pedigrees as a tool to control the rates of inbreeding and maintain genetic diversity in sustainable breeding programs. Genetic studies for paternity testing have so far been carried out using microsatellite markers developed from other species of the same genus [4,20]. Using heterologous microsatellites can introduce biases when assessing population variability; for instance, null alleles and homoplasy may falsify the genetic structure and parentage testing.

The aim of this study is to develop a novel genotyping-by-sequencing (GBS) marker panel for parentage testing using single nucleotide polymorphism (SNP) data obtained from a genotyping array and a whole *S. lalandi* genome assembly, as presented elsewhere [21]. We focused our attention on the performance of this panel under real experimental conditions, considering the information of progeny data from two source populations, one native to Chile and another to New Zealand (NZ). This information was used to calculate predicted inbreeding rates using the genetic contribution theory and variation in family size. The results are discussed from the perspective of developing breeding programs for "these new" aquaculture species that are effective in increasing the genetic gain while constraining the rates of inbreeding.

2. Materials and Methods

2.1. Production System and Data Sampling

Fish from Chile were sampled from a captive *S. lalandi* broodstock population held in Acuinor S.A. Company's facilities in Caldera City, located in the Atacama Region. These fish were captured from a wild population at Punta Frodden as well as from different locations near Caldera (27.0° S; 70.8° W) and kept to conform to the base population of the national *S. lalandi* breeding programs (56 dams and 51 sires). The individuals were arranged in four independent breeding units (R1, R2, R3, and R4 with about 20–30 fish per tank). Hence, they were exposed to different photoperiods and temperature increases to ensure the availability of larvae throughout the entire year. The progeny was reared under standard farm conditions, and the fish were kept until harvest at about 3 Kg in different Recirculating Aquaculture Systems (RAS) units. We sampled fin clips from all Chilean broodstock and 161 randomly sampled progeny from the different breeding units (R1, R2, and R3) produced by mass spawnings. All fish were anaesthetized before sampling using MS-222 as part of the management plan to measure production variables.

Fish from New Zealand ($n = 31$) were sampled from a captive broodstock originated from the east coast of the Northland province (35° S, 174° E) North Island and was kept in captivity for 2 to 8 years. The animals had been held at the NIWA Northland Aquaculture Centre, located in Ruakaka.

Fin clips were kept frozen or in >70% ethanol before DNA extraction. Genomic DNA was extracted using the NucleoSpin® Plant II kit (Macherey-Nagel®, Düren, Germany) according to the manufacturer's instructions. The samples were quantified using a Qubit fluorometer (Thermo Fisher Scientific, Waltham, MA, USA) with the Qubit dsDNA BR Assay Kit (Thermo Fisher Scientific, Waltham, MA, USA). DNA was normalized to a final concentration of 2 ng/µL.

2.2. Genotyping-By-Sequencing (GBS) Panel Construction

The marker panel was developed using the genomic resources within the national breeding program of *S. lalandi*. To develop the genetic resources needed for its implementation, we first produced an in-house genome assembly for the species. This reference was developed by sequencing a single male with a coverage of 70× using a mixture of single and mate Illumina pair-ends reads. The draft genome was assembled using MaSuRCA [22] with the default parameters. An additional 34 individuals (17 males and 17 females) were sequenced at 10× coverage to discover SNPs using the scaffolds obtained by MASURCA and the procedures explained below. These fish were also used to discover SNPs associated with the sex determination gene, that appeared to be causal [23]. We used 90 K SNPS selected (out of more than 5 million SNPs discovered) to develop the 90 K-SNP genotyping array (constructed by Thermofisher™, Waltham, MA, USA). Scaffolds were anchored to linkage groups with CHROMONOMER [24] using the linkage distance between 90 K SNPs markers as obtained from the genotyping array using LEP-MAP (we used information on 200 progeny from 10 full sibs families to generate the linkage map [25]). The assembled genome comprised 24 linkage groups with a total genome size of about 657 Mb encompassing 95% of the total expected genome. More detailed information will be provided in a separate study (in preparation [23,26]).

Selected markers used for the construction of the GBS panel were obtained from genotype data of the base (founder) population of the national Chilean breeding programme of *S. lalandi*. In this case, a total of 300 SNPs were selected using genotype data obtained from the 90 K-SNP-genotyping array of the *S. lalandi* broodstock population [23] as obtained from data from the broodstock population genotyped with the 90 K-SNP-genotyping array. The SNPs (280 markers) used for parentage assignment in the panel were evenly placed across all 24 linkage groups. The markers were selected based on informativeness using a minimum allele frequency (MAF) of 0.34. The average marker spacing is 2 Mb (the minimum and maximum distances between pairs of markers were 0.7 Mb and 13 Mb, respectively; see Figure 1). All the markers selected for the panel followed Hardy-Weinberg equilibrium in the broodstock population and showed no linkage disequilibrium within linkage groups. We included 20 markers in the vicinity of the diagnostic SNP used for sex prediction (these markers were obtained from a genome-wide association analysis as explained above and will be published in a separate study [23,26]).

For obtaining the actual genotypes of the progeny, a targeted GBS protocol was used. A total of 300 primer pairs were developed based on proprietary software from Thermo Fisher (https://www.thermofisher.com/cl/es/home/global/forms/agriseq-breeding.html (accessed on 20 January 2021); Thermo Fisher Scientific, Waltham, MA, USA). A total of 192 libraries were prepared using AgriSeq™ HTS Library Kit (Thermo Fisher Scientific). Libraries were genotyped using the Ion 540 Chef kit along with the Ion 540 Chip (~80 million reads) kit (Thermo Fisher Scientific). Sequences were mapped to the *S. lalandi* genome assembly using BWA [27]. SAM files were sorted with SAMtools [28], and PCR duplicates were removed using SAMBAMBA [29]. SNPs were identified using FreeBayes (https://arxiv.org/abs/1207.3907v2 accessed on 30 March 2021) with default settings. The initial set of SNPs identified was filtered using vcftools [30] based on the following criteria (derived from the FreeBayes output): (1) a Phred-scaled SNP quality score with significance greater than 30; (2) minimum allele frequency of 0.49; (3) maximum percentage of missing values of 0.20; and (4) the maximum number of alleles as 2.

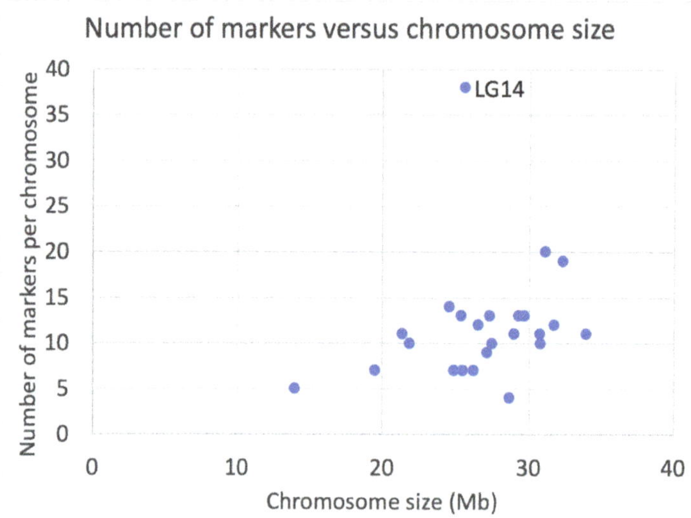

(**a**)

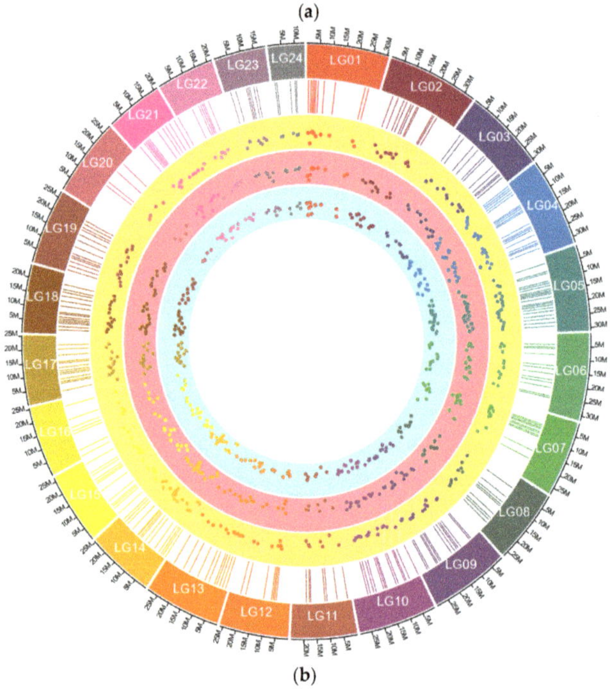

(**b**)

Figure 1. (**a**) Number of markers per chromosome (Y-axis) versus chromosome (linkage group: LG) size in Mb (assume 1cM equals Mb: X-axis). The linkage group 14 (LG14) included the markers for sex determination. (**b**) Markers used for the GBS panel and MAF. The outer layer shows the genome, the second layer (rectangle) shows the position of the markers, while the third, fourth, and fifth layers the observed MAF for the New Zealand and Chilean populations.

2.3. Genotype Detection and Parentage Analysis Using the GBS Panel

SNP variation statistics were obtained separately for the different datasets (New Zealand (n = 31) and Chile (n = 161)) when using the GBS panel. We calculated minimum allele frequency (MAF), Hardy-Weinberg χ^2 statistics, and observed heterozygosity (He) using PLINK [31]. The cumulative probability of exclusion (CPE) for single-parent (CPE-1) and both-parent inference (CPE-2) were calculated using formulae of Jamieson and Taylor [32], as well as the polymorphic information content (PIC) with equations obtained from [33].

Parentage assignment was carried out using data from the Chilean broodstock (which was genotyped by the 90 K genotyping array) and progeny (161, as explained above) with AlphaAssign [34]. This procedure relies on a maximum likelihood approach to infer parents (sires or dams) using default parameters. In practice, sex is predicted at the progeny level with the GBS marker panel, making it possible to independently assign sires or dams in the next generation when selecting potential broodstock from the progeny available. This is expected to increase the accuracy of the procedure by reducing the number of possible single-parent (sire or dam) and progeny pairs when calculating the likelihood [35]. Since we did not use the GBS panel for genotyping the parents, we obtained genotypes of the parental generation by extracting the markers selected for the GBS panel from data obtained from genotypes of broodstock using the 90 K-SNP-genotyping array.

The predicted rates of inbreeding were calculated using two different methods. The first one uses genetic contribution theory, assuming random selection and discrete generations [36]. In this case, an estimate can be obtained by using the following approximation (Equation (1)):

$$\Delta F = \frac{1}{4}\left[\sum r_i^2\right] \tag{1}$$

where r_i is the mean parental genetic contribution for each parent calculated, from generation 0 to generation 1, using the predicted pedigree. The mean genetic contributions were obtained from the additive relationship values between parents and progeny and then averaged over the total number of progenies. The values of r_i summed over each sex (dam or sires) equal 0.5.

We also used the method derived by [37] for estimating inbreeding effective size that incorporates the offspring contribution (Equation (6), without selfing from [37]), as:

$$Ne = \frac{2S - 2}{\frac{\sum (k_i^2)}{2S} - 1} \tag{2}$$

where S is the total number of progenies, and k_i is the family size for each of the parents (males and females). This method is subject to relatively large standard errors when the number of parents is large and the number of offspring is small (giving an upward bias to Ne, as obtained from simulations [37]). The estimates of the rates of inbreeding and effective size were obtained as follows:

$$\Delta F = \frac{1}{2Ne} \tag{3}$$

The effective numbers of founders (ENF), was calculated as [38]:

$$ENF = 1/\left[\sum r_i\right] \tag{4}$$

All these calculations were carried out using PEDIG [38] and Excel spreadsheets, using the pedigree predicted with AlphaAssign. Estimates of Ne and predicted rates of inbreeding were obtained using information from all the breeding units together as well as separately.

3. Results

3.1. Genotyping-By-Sequencing (GBS) Marker Panel Assessment

We assessed the informativeness of the GBS marker panel across the different populations analyzed (New Zealand and Chile). After sequencing, a total of 188 out of 192 samples passed with a minimum sample call rate of 97% (with an average of 98%). The average coverage was 1152× per marker. Seven SNP markers did not pass the quality control since they had a low call rate and were excluded. One additional marker showed a MAF equal to 0 in the progeny. Therefore, a total of 272 markers were mapped to the 24 chromosomes using BWA (Supplementary Table S1; Figure 1a). These markers were used in the final parentage analysis based on the GBS panel, and 20 markers were used for sex prediction. The list of SNPs and their detailed information is given in Supplementary Table S1.

When examining the 272 markers used for paternity testing, MAF values ranged from 0.11 to 0.50, with an average of 0.36 for the entire dataset (Figure 1b). The percentage of SNPs with a MAF between 0.40 and 0.50 was 36% and 40%, respectively, for the Chilean and New Zealand populations. In addition, the average polymorphic information content (PIC) was 0.50 and 0.53 for the Chilean and New Zealand populations, respectively. The realized average linkage disequilibrium (r^2) between adjacent marker pairs within chromosomes was 0.05 and 0.04 for the Chilean and New Zealand populations, respectively. In addition, the CPE for single-parent (CPE-1) and both-parent (CPE-2) inference were in all the populations higher than 0.999.

3.2. Parentage Testing and Distribution of Genetic Contributions

The average likelihood difference between unrelated parents (sires or dams), which is calculated to assign parentage, was substantially higher than in the case of chosen candidate parents using all marker data (these values were calculated with AlphaAssign using default parameters, Figure 2). The parents that were not assigned to offspring had values for the likelihood difference near zero. We tested individuals from New Zealand (as progeny) with putative parents from Chile. In this analysis, no putative sires or dams were assigned to the individuals coming from the New Zealand sample population, so the rate of false positives was negligible (data not shown).

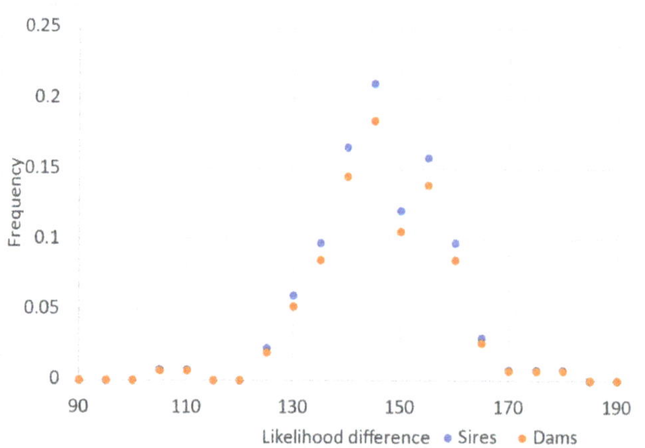

Figure 2. Histogram of the likelihood difference between the chosen candidate (sires or dams) and unrelated sires or dams for each progeny evaluated using all markers included in the panel (272 SNPs).

We tested different scenarios to assess the effect of the number of markers and the performance of the maximum likelihood approach for assessing parentage (Figure 3). This was achieved by randomly deleting markers used to infer parentage with PLINK using the thin option. We found a linear trend between the number of markers and the difference between the likelihood for a specific set of markers and the full set of markers (the ratio of the average difference between the likelihood for a reduced set of markers and the full set of markers was 20 to 100%). The linear regression coefficient was about 0.04 units per marker (Figure 3a). The probability of non-assignment (using the number of parents not assigned due to a reduced likelihood difference from the total number of parents assigned) is highly dependent on the number of markers. The threshold for accurately assigning parental data was about 150 markers. (See Figure 3b, in which the value of the proportion of non-assignment remains at a maximum value of 5% after 150 markers).

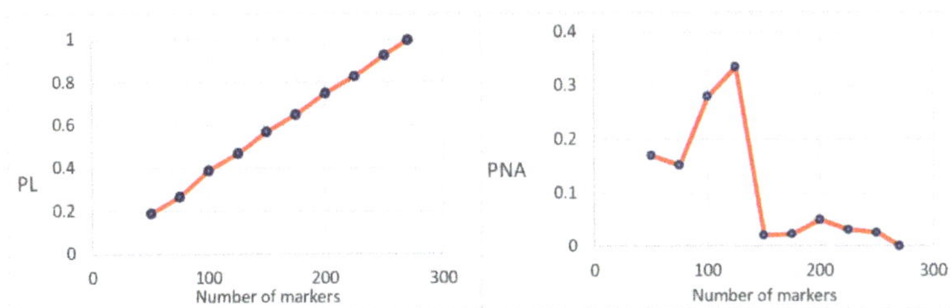

(a) (b)

Figure 3. (**a**) Ratio of the average difference between the likelihood (PL) of the chosen parent and an unrelated individual for a different number of markers (expressed as a proportion when using all markers); (**b**) Proportion of non-assignment (PNA) given a different number of markers used for parentage testing.

The parental assignment revealed an extreme asymmetry in the parental contributions, particularly concerning the females (Figure 4). In the more extreme case, a dam was indicated as the single parent of almost all the progeny in one of the production batches (see Figure 4 and Table 1). The mean number of offspring per dam (17 out of 55 dams were assigned) was 12, with a very high variance (293). The males contributed more evenly to the gene pool. The average offspring count of males (36 out of 51 males were assigned) was 6, but with a much lower variance (24).

(a) (b)

Figure 4. Bubble graphs showing the mean genetic contributions of single dams (**a**) and sires (**b**) across different breeding units. The Y-axis denotes the proportion of offspring per dam and sire.

Table 1. Predicted rates of inbreeding and effective population size (*Ne*) for different production batches. DF_1 (Predicted rate of inbreeding: Wooliams and Thompson, 1991); Ne_1 (*Ne*: Wooliams and Thompson, 1991); DF_2, (Predicted rate of inbreeding: Waples and Waples, 2005); Ne_2 (Waples and Waples, 2005); ONF (Observed number of females); ON_M (Observed number of males); N_F (Number of contributing females); N_M (Number of contributing males); ENF, (Effective number of founders); E_{kf} (Average family size of females); E_{km} (Average family size of males); V_{kf} (Variance of family size of females); V_{km} (Variance of family size of males). R3′ corresponds to a different production batch from the breeding unit R3.

Breeding Unit	ΔF1	Ne_1	ΔF2	Ne_2	ONF	ONM	NCF	NCM	ENF	E_{kf}	E_{km}	V_{kf}	V_{km}
R1	0.06	7.9	0.12	4.1	18	13	5	9	4	7	4	192	17
R2	0.06	8.6	0.09	5.8	15	9	5	10	4.9	12	3	292	6
R3	0.07	7.4	0.11	4.5	13	15	3	6	3.7	12	4	96	1
R3′	0.04	12.7	0.06	9.2	13	15	7	14	7.1	9	3	115	21
Overall	0.02	23.8	0.03	15.1	46	35	17	41	13.3	12	6	293	25

The predicted rates of inbreeding varied significantly between methods, especially when considering individual breeding units. Overall, it was lower when using the method using genetic contributions (Table 1). Nonetheless, the predicted rates of inbreeding were in general very high when considering specific production groups, which is not surprising given the very low numbers of parents contributing within each batch, ranging between ΔF_1 4% and 7% and ΔF_2 ranging between 6% and 12% (Table 1). When considering all the groups together, the predicted rate of inbreeding was about 2.1% and 3.3% for DF1 and DF2, respectively. This gave the values of the effective population sizes of about 23.8 and 15.1 for Ne_1 and Ne_2 (see Table 1). The effective number of founders was about one-third the number of breeders contributing progeny in the next generation (Table 1).

4. Discussion

Molecular parentage testing in aquaculture has been traditionally carried out using microsatellites [20]. However, the general lack of species-specific standardized panels prevents efforts to automate the process, resulting in microsatellites not being routinely used for parentage analyses as a tool to control inbreeding. Furthermore, many microsatellites developed on congeners (e.g., *S. quinqueradiata*) are not usable due to their low informativeness and the presence of null alleles. Indeed, our research team showed that 10 out of 25 non-focal microsatellites are useless for parentage analysis [20]. Additionally, microsatellite panels are not useful for sex determination, requiring a separate step of high-resolution melt PCR [21]. For all these reasons, it is important to develop marker panels using SNPs, which are faster to obtain, more accurate, and cheaper when compared with microsatellites.

In this study, the realized value for the cumulative probability of exclusion for a single or both parents was 1, suggesting that the GBS marker panel of 272 markers can be used very efficiently for parentage testing in captive *S. lalandi* populations. In fact, when using only 150 SNPs as markers, there is enough power to perform parentage analysis (Figure 3), with a rate of non-assignment lower than 5%. We also found that false positives are rare, since when performing parentage testing on an unrelated population, no related parents were obtained when more than 150 markers were used.

4.1. Predicted Rates of Inbreeding in S. lalandi

The predicted rates of inbreeding calculated in this study should be interpreted as long-term estimates and not as forecasts of average inbreeding in the next generation. Furthermore, these rates should be thought of as rough estimates as they only rely on information from a relatively low number of single spawning events over limited time spans. Therefore, the genetic contributions are not at their asymptotic values, leading to an underestimation of the rates of inbreeding in the long term. This can be seen in

Table 1, where values based on genetic contributions are smaller than estimates from another study [37].

Some unintentional selection is likely ongoing in the population since, in large production batches, individuals are graded to optimize feeding practices while decreasing the rates of cannibalism [39]. This preselection phase may explain, to a certain degree, the high variance of some of the parents assigned. Nevertheless, specific females seem to dominate single spawning events since they appear in half-sib maternal families with several males, which is in accordance with what had previously been observed in this species [40]. Therefore, inbreeding is of concern in this species since, without intervention, it will be extremely high within a short time, and strategies to control its rates are needed.

Overall, the predicted inbreeding rates in our breeding units were generally very high. In all cases, values of the effective population size were much smaller than the ones required to reduce the risk of extinction in species with high reproductive output [14,35,41]. In practice, a cumulative number of spawnings should be used in order to secure a sustainable breeding program (see below). This means that the generation interval will be higher when compared with other species (i.e., salmonids), since multiple egg batches should be kept over longer time spans.

There is some scope to constrain the rates of inbreeding by avoiding mating between related individuals, at least in the short term. In our case, since the founders of the population are the actual broodstock used for parentage testing, it is straightforward to select replacements based on the minimum inbreeding (using the progeny candidates) using a rotational mating scheme. This system has been devised in the classical paper by Kimura & Crow [42] and applied to Coho salmon (*Oncorhynchus kisutch*) to replace broodstock [43]. Nevertheless, this methodology would only be useful for the short term, and a high rate of inbreeding has been observed in the long term [14]. In this mating system, broodstock was kept in a few separate groups, and males were transferred sequentially and circularly between neighboring groups, which were kept isolated [44]. In the simplest system to be considered, n males and n females were arranged alternately so that each potential sire or dam mated with individuals of a neighboring reproduction unit. In this classical system, the second generation is the product of a half-sib or full-sib mating (when dams are also replaced). We have applied this replacement method by sequentially assigning sires and dams to unrelated reproduction units (R1, R2, R3, and R4). This procedure gave null inbreeding in the second generation, as expected since the co-ancestry between breeders within tanks was still 0 (only unrelated founders are available in generation 0). Further investigation is needed to understand the consequences of this type of mating in the long term.

4.2. Low-Density Marker Panels, Genomic Prediction, and Two-Stage Selection Programmes

We have shown that specific progeny batches represent a reduced number of founders from the previous generation, even though the number of progenies can be large (since potentially millions of eggs are produced in each spawning event). This situation will lead to a series of practical problems when implementing genomic breeding programs in practice.

First, if a reduced number of batches are produced in each generation, it will be difficult to predict accurate breeding values using SNPs since the associations between markers and quantitative trait loci (QTL) would not represent associations at the population level (most of the information will come from a small number of full-sib families). Secondly, progeny from single spawning events are related to half-sib maternal families, and therefore (when mated), they will produce inbred offspring. In the case of *S. lalandi*, several batches are produced within a year, but the progeny of each batch [30] mostly originates from a single dominant female (Figure 4 and [20,40]). For these reasons, it is important to include selection candidates from as large a number of batches as practicable in order to represent a significant proportion of the genomic variation at the population level.

Secondly, one major issue in developing breeding programs for *S. lalandi* using genomic information is the genotyping cost of using, in particular, SNP-genotyping arrays. The price of the GBS marker panel is about 10% of the cost of the full genotyping array, giving some scope for implementing profitable selection programs, as we detailed as follows.

We have previously found that in salmonids, a two-stage selection program gave the most profitable breeding results at the expense of maximum genetic gains and constrained rates of inbreeding [45]. Therefore, an essential part of the design should follow a procedure for maintaining high levels of genetic variability by keeping a larger number of production batches in the first stage of selection while selecting for the traits of interest in different stages of the life cycle. This can be completed in the early stages, when the fish are still comparatively small in size and, hence, easier to manage (when fish can be handled in large batches from different spawnings at lower body weights, for example, at 10 months). At this stage, a set of diverse production batches can be sampled based on the trait of interest and then subjected to paternity testing for culling related individuals, based on a procedure minimizing coancestry levels between chosen candidates. The parentage testing using the GBS marker panel in this stage is essential to reconstruct the pedigree of the population and secure enough individuals of each sex differing by some degree of sexual dimorphism for body weight (females appear to be heavier than males during maturation) [46].

A second stage should be performed by selecting individuals based on the genomic breeding values of the selected population while applying optimal contribution selection [47]. In this case, breeding values should be predicted for harvest weight (usually 3–5 kg.) using information from a reference population and a denser genotyping array or a modification of a GBS genotyping incorporating SNP associated with traits of interest or using imputation. The exact proportions of individuals to be used for either stage require further investigation, as does the assessment of the optimum selection intensity to maximize profit [45,48].

Another possibility is to use a smaller number of markers in genotyping panels for selecting individuals in a single stage (usually 3–5 kg) using imputation. This will decrease the genotyping costs while possibly maintaining the rates of gain, as has been suggested before [49]. Nonetheless, it is difficult to constrain the inbreeding rate in these scenarios while using a reduced number of production batches for selecting individuals as potential selected broodstock candidates. Further research on this subject is required to disentangle the factors enabling sustainable rates of genetic gain.

5. Conclusions

In this study, we developed a marker panel to perform parentage testing as well as sex determination ("Martínez manuscript in prep.") in *S. lalandi*. This resource is essential when implementing breeding programs, since controlling inbreeding rates requires not only parentage testing but also predicting candidate sex. We used markers spanning all the linkage groups devised using recombination data from a reference genome. This enabled us to select markers with high variability and minimal linkage disequilibrium. The validation in two unrelated populations (New Zealand and Chile) suggests that only half of the markers are required for parentage testing with sufficient power. No false positives were detected when considering cross-parentage testing between populations. This panel is key to developing sound breeding strategies that constrain the rates of inbreeding in the short and long term. This is particularly important considering the difficulties in carrying out breeding programs in species subjected to communal rearing and uncontrolled reproduction. These achievements will increase the viability of this new aquaculture resource and, as such, the success of the associated business activities.

Supplementary Materials: The following supporting information can be downloaded at: https://www.mdpi.com/article/10.3390/ani13050913/s1, Table S1: Position and Minimum allele frequency for each SNP used in paternity testing. The position was obtained by using [26].

Author Contributions: Conceptualization, V.M.; methodology, V.M.; validation, V.M. and N.G.; formal analysis, V.M.; investigation, V.M., N.G. and A.S.; resources, A.S.; data curation, V.M.; writing—original draft preparation, V.M.; writing—review and editing, V.M., N.G. and A.S; project administration, V.M.; funding acquisition, V.M. All authors have read and agreed to the published version of the manuscript.

Funding: This research was funded by FONDECYT grant N° 1191189, from ANID, Ministry of Science, Chile, and by the ENLACE N° 1074/2022 grant from the University of Chile and by the Strategic Science Investment Fund of the NZ Ministry of Business, Innovation and Employment.

Institutional Review Board Statement: All procedures involving the handling and treatment of animals were approved by the Bioethics Committee of the Veterinary Science Faculty, University of Chile, Chile (Permit Number. 10-2015).

Informed Consent Statement: Not applicable.

Data Availability Statement: The marker data is available upon request to the corresponding author. All the primers used for sequencing will be available for sequencing using Ion 540 chef upon request. The reference genome will be available upon request.

Conflicts of Interest: The authors declare no conflict of interest. The funders had no role in the design of the study; in the collection, analyses, or interpretation of data; in the writing of the manuscript, or in the decision to publish the results.

References

1. Nugroho, E. Population Genetic Studies on the Aquaculture Fish in Genus *Seriola* for Their Risk Management. Ph.D. Thesis, Northeastern University, Shenyang, China, 2001.
2. Randall, J.E.; Lim, K.K.P. A Checklist of the Fishes of the South China Sea. *Raffles Bull. Zool.* **2000**, *8*, 569–667.
3. Lovatelli, A.; Holthus, P.F.; Food and Agriculture Organization of the United Nations (Eds.) *Capture-Based Aquaculture: Global Overview*; FAO Fisheries Technical Paper; Food and Agriculture Organization of the United Nations: Rome, Italy, 2008; ISBN 978-92-5-106030-8.
4. Fernández, G.; Cichero, D.; Patel, A.; Martínez, V. Genetic Structure of Chilean Populations of *Seriola lalandi* for the Diversification of the National Aquaculture in the North of Chile. *Lat. Am. J. Aquat. Res.* **2015**, *43*, 374–379. [CrossRef]
5. Eschmeyer, W.N.; Herald, E.S. *A Field Guide to Pacific Coast Fishes: North America*; Houghton Mifflin Harcourt: Boston, MA, USA, 1999; ISBN 978-0-618-00212-2.
6. Dyer, B.S.; Westneat, M.W. Of Juan Fernández Archipelago and Desventuradas Islands, Chile. *Rev. Biol. Mar. Oceanogr.* **2010**, *1*, 589–617. [CrossRef]
7. Risk Assessment for Metazoan Parasites of Yellowtail Kingfish *Seriola lalandi* (Perciformes: Carangidae) in South Australian Sea-Cage Aquaculture-ScienceDirect. Available online: https://www.sciencedirect.com/science/article/pii/S0044848607002761?casa_token=KywUG1A7EhQAAAAA:iNT1wbiCqiFrCcll-CcXqEtVgDOQRCXEQ-bz2N5nm8M2_CzLC3ct8U-isFyOj1sbv5BZRFPdMG0C (accessed on 22 September 2022).
8. Symonds, J.; Walker, S.; Pether, S.; Gublin, Y.; McQueen, D.; King, A.; Irvine, G.; Setiawan, A.; Forsythe, J.; Bruce, M. Developing Yellowtail Kingfish (*Seriola lalandi*) and Hāpuku (*Polyprion oxygeneios*) for New Zealand Aquaculture. *N. Z. J. Mar. Freshw. Res.* **2014**, *48*, 371–384. [CrossRef]
9. Kingfish Company Obtains Federal Permit for Maine RAS Facility. Available online: https://www.rastechmagazine.com/kingfish-company-obtains-federal-permit-for-maine-ras-facility/ (accessed on 23 September 2022).
10. Growing RAS in Chile. Available online: https://www.rastechmagazine.com/growing-ras-in-chile/ (accessed on 23 September 2022).
11. Gjedrem, T. Genetic Improvement of Cold-Water Fish Species. *Aquac. Res.* **2000**, *31*, 25–33. [CrossRef]
12. Aho, T.; Rönn, J.; Piironen, J.; Björklund, M. Impacts of Effective Population Size on Genetic Diversity in Hatchery Reared Brown Trout (*Salmo trutta* L.) Populations. *Aquaculture* **2006**, *253*, 244–248. [CrossRef]
13. Eknath, A.E.; Doyle, R.W. Effective Population Size and Rate of Inbreeding in Aquaculture of Indian Major Carps. *Aquaculture* **1990**, *85*, 293–305. [CrossRef]
14. Martinez, V.; Dettleff, P.J.; Galarce, N.; Bravo, C.; Dorner, J.; Iwamoto, R.N.; Naish, K. Estimates of Effective Population Size in Commercial and Hatchery Strains of Coho Salmon (*Oncorhynchus kisutch* (Walbaum, 1792)). *Animals* **2022**, *12*, 647. [CrossRef]
15. Morvezen, R.; Boudry, P.; Laroche, J.; Charrier, G. Stock Enhancement or Sea Ranching? Insights from Monitoring the Genetic Diversity, Relatedness and Effective Population Size in a Seeded Great Scallop Population (*Pecten maximus*). *Heredity* **2016**, *117*, 142–148. [CrossRef]
16. Saura, M.; Caballero, A.; Santiago, E.; Fernández, A.; Morales-González, E.; Fernández, J.; Cabaleiro, S.; Millán, A.; Martínez, P.; Palaiokostas, C.; et al. Estimates of Recent and Historical Effective Population Size in Turbot, Seabream, Seabass and Carp Selective Breeding Programmes. *Genet. Sel. Evol.* **2021**, *53*, 85. [CrossRef]

17. Vela Avitúa, S.; Montaldo, H.H.; Márquez Valdelamar, L.; Campos Montes, G.R.; Castillo Juárez, H. Decline of Genetic Variability in a Captive Population of Pacific White Shrimp *Penaeus* (*Litopenaeus*) *vannamei* Using Microsatellite and Pedigree Information. *Electron. J. Biotechnol.* **2013**, *16*, 9. [CrossRef]
18. Nguyen, N.H.; Whatmore, P.; Miller, A.; Knibb, W. Quantitative Genetic Properties of Four Measures of Deformity in Yellowtail Kingfish *Seriola lalandi* Valenciennes, 1833. *J. Fish Dis.* **2016**, *39*, 217–228. [CrossRef] [PubMed]
19. Whatmore, P.; Nguyen, N.H.; Miller, A.; Lamont, R.; Powell, D.; D'Antignana, T.; Bubner, E.; Elizur, A.; Knibb, W. Genetic Parameters for Economically Important Traits in Yellowtail Kingfish *Seriola lalandi*. *Aquaculture* **2013**, *400–401*, 77–84. [CrossRef]
20. Dettleff, P.; Hernandez, E.; Partridge, G.; Lafarga-De la Cruz, F.; Martinez, V. Understanding the Population Structure and Reproductive Behavior of Hatchery-Produced Yellowtail Kingfish (*Seriola lalandi*). *Aquaculture* **2020**, *522*, 734948. [CrossRef]
21. Martinez, V. Genome wide association of sex dtermibnation in Seriola lalandi. *in preparation*.
22. Zimin, A.V.; Puiu, D.; Luo, M.-C.; Zhu, T.; Koren, S.; Marçais, G.; Yorke, J.A.; Dvořák, J.; Salzberg, S.L. Hybrid Assembly of the Large and Highly Repetitive Genome of *Aegilops tauschii*, a Progenitor of Bread Wheat, with the MaSuRCA Mega-Reads Algorithm. *Genome Res.* **2017**, *27*, 787–792. [CrossRef]
23. Martinez, V. Diversification of Marine Aquaculture through Genomics: Applications to Breeding Programs in Novel Species. In Proceedings of the Plant and Animal Genome XXVII, San Diego, CA, USA, 14 January 2019; Available online: https://www.youtube.com/watch?v=oIGf6-3ExVc (accessed on 10 March 2021).
24. Catchen, J.; Amores, A.; Bassham, S. Chromonomer: A Tool Set for Repairing and Enhancing Assembled Genomes Through Integration of Genetic Maps and Conserved Synteny. *G3 (Bethesda)* **2020**, *10*, 4115–4128. [CrossRef]
25. Rastas, P. Lep-MAP3: Robust Linkage Mapping Even for Low-Coverage Whole Genome Sequencing Data. *Bioinformatics* **2017**, *33*, 3726–3732. [CrossRef]
26. Martinez, V.; Galarce, N. 023 The Reference Genome of *Seriola lalandi* Gave Insights on Sex Determination and Major Histocompatibility Variation. 2023; *in preparation*.
27. Li, H.; Durbin, R. Fast and Accurate Short Read Alignment with Burrows-Wheeler Transform. *Bioinformatics* **2009**, *25*, 1754–1760. [CrossRef]
28. Li, H.; Handsaker, B.; Wysoker, A.; Fennell, T.; Ruan, J.; Homer, N.; Marth, G.; Abecasis, G.; Durbin, R.; 1000 Genome Project Data Processing Subgroup. The Sequence Alignment/Map Format and SAMtools. *Bioinformatics* **2009**, *25*, 2078–2079. [CrossRef]
29. Tarasov, A.; Vilella, A.J.; Cuppen, E.; Nijman, I.J.; Prins, P. Sambamba: Fast Processing of NGS Alignment Formats. *Bioinformatics* **2015**, *31*, 2032–2034. [CrossRef]
30. Danecek, P.; Auton, A.; Abecasis, G.; Albers, C.A.; Banks, E.; DePristo, M.A.; Handsaker, R.E.; Lunter, G.; Marth, G.T.; Sherry, S.T.; et al. The Variant Call Format and VCFtools. *Bioinformatics* **2011**, *27*, 2156–2158. [CrossRef] [PubMed]
31. Chang, C.C.; Chow, C.C.; Tellier, L.C.; Vattikuti, S.; Purcell, S.M.; Lee, J.J. Second-Generation PLINK: Rising to the Challenge of Larger and Richer Datasets. *GigaScience* **2015**, *4*, s13742-015. [CrossRef] [PubMed]
32. Jamieson, A.; C S Taylor, S. Comparisons of Three Probability Formulae for Parentage Exclusion. *Anim. Genet.* **1997**, *28*, 397–400. [CrossRef] [PubMed]
33. Botstein, D.; White, R.L.; Skolnick, M.; Davis, R.W. Construction of a Genetic Linkage Map in Man Using Restriction Fragment Length Polymorphisms. *Am. J. Hum. Genet.* **1980**, *32*, 314–331.
34. Whalen, A.; Gorjanc, G.; Hickey, J.M. Parentage Assignment with Genotyping-by-Sequencing Data. *J. Anim. Breed. Genet.* **2019**, *136*, 102–112. [CrossRef]
35. Martinez, V. Marker-Assisted Selection in Fish and Shellfish Breeding Schemes. In *Marker-Assisted Selection: Current Status and Future Perspectives in Crops, Livestock, Forestry and Fish*; Guimaraes, E.P., Ruane, J., Scherf, B.D., Sonnino, A., Dargie, J.D., Eds.; Food and Agriculture Organization of the United Nations: Rome, Italy, 2007; 494p.
36. Woolliams, J.A.; Berg, P.; Dagnachew, B.S.; Meuwissen, T.H.E. Genetic Contributions and Their Optimization. *J. Anim. Breed. Genet.* **2015**, *132*, 89–99. [CrossRef]
37. Waples, R.S.; Waples, R.K. Inbreeding Effective Population Size and Parentage Analysis without Parents. *Mol. Ecol. Resour.* **2011**, *11* (Suppl. 1), 162–171. [CrossRef]
38. Boichard, D.; Maignel, L.; Verrier, É. The Value of Using Probabilities of Gene Origin to Measure Genetic Variability in a Population. *Genet. Sel. Evol.* **1997**, *29*, 5. [CrossRef]
39. Hauville, M.; Zambonino-Infante, J.; Migaud, H.; Bell, J.G.B.; Main, K. Effects of Probiotics on Pompano (*Trachinotus carolinus*), Common Snook (*Centropomus undecimalis*), and Red Drum (*Sciaenops ocellatus*) Larvae. *Commun. Agric. Appl. Biol. Sci.* **2013**, *78*, 180–183.
40. Schmidt, E.; Stuart, K.; Hyde, J.; Purcell, C.; Drawbridge, M. Spawning Dynamics and Egg Production Characteristics of Captive *Seriola dorsalis* Assessed Using Parentage Analyses. *Aquac. Res.* **2021**, *52*, 4050–4063. [CrossRef]
41. Caballero, A.; Bravo, I.; Wang, J. Inbreeding Load and Purging: Implications for the Short-Term Survival and the Conservation Management of Small Populations. *Heredity* **2017**, *118*, 177–185. [CrossRef]
42. Kimura, M.; Crow, J.F. On the Maximum Avoidance of Inbreeding. *Genet. Res.* **1963**, *4*, 399–415. [CrossRef]
43. Hershberger, W.K.; Myers, J.M.; Iwamoto, R.N.; Mcauley, W.C.; Saxton, A.M. Genetic Changes in the Growth of Coho Salmon (*Oncorhynchus kisutch*) in Marine Net-Pens, Produced by Ten Years of Selection. *Aquaculture* **1990**, *85*, 187–197. [CrossRef]
44. Nomura, T.; Yonezawa, K. A Comparison of Four Systems of Group Mating for Avoiding Inbreeding. *Genet. Sel. Evol.* **1996**, *28*, 141. [CrossRef]

45. Martinez, V.; Kause, A.; Mäntysaari, E.; Mäki-Tanila, A. The Use of Alternative Breeding Schemes to Enhance Genetic Improvement in Rainbow Trout: II. Two-Stage Selection. *Aquaculture* **2006**, *254*, 195–202. [CrossRef]
46. Poortenaar, C.W.; Hooker, S.H.; Sharp, N. Assessment of Yellowtail Kingfish (*Seriola lalandi lalandi*) Reproductive Physiology, as a Basis for Aquaculture Development. *Aquaculture* **2001**, *201*, 271–286. [CrossRef]
47. Sonesson, A.K.; Woolliams, J.A.; Meuwissen, T.H. Genomic Selection Requires Genomic Control of Inbreeding. *Genet. Sel. Evol.* **2012**, *44*, 27. [CrossRef] [PubMed]
48. Jibrila, I.; Ten Napel, J.; Vandenplas, J.; Veerkamp, R.F.; Calus, M.P.L. Investigating the Impact of Preselection on Subsequent Single-Step Genomic BLUP Evaluation of Preselected Animals. *Genet. Sel. Evol.* **2020**, *52*, 42. [CrossRef] [PubMed]
49. Kriaridou, C.; Tsairidou, S.; Houston, R.D.; Robledo, D. Genomic Prediction Using Low Density Marker Panels in Aquaculture: Performance Across Species, Traits, and Genotyping Platforms. *Front. Genet.* **2020**, *11*, 124. [CrossRef] [PubMed]

animals

MDPI

Article

Assessing Molecular Diversity in Native and Introduced Populations of Red Wood Ant *Formica paralugubris*

Alberto Masoni [1,*], Andrea Coppi [1], Paride Balzani [1,2], Filippo Frizzi [1], Renato Fani [1], Marco Zaccaroni [1] and Giacomo Santini [1]

1 Department of Biology, University of Florence, 50019 Florence, Italy
2 Faculty of Fisheries and Protection of Waters, South Bohemian Research Center of Aquaculture and Biodiversity of Hydrocenoses, University of South Bohemia in České Budějovice, 38925 Vodňany, Czech Republic
* Correspondence: alberto.masoni@unifi.it

Simple Summary: Red wood ants are ecologically dominant ant species that play key roles in boreal forest ecosystems, where they greatly influence the habitat dynamics with their predatory activity. During the last century, they were largely employed as biocontrol agents in Italy against forest pests, and thousands of nests were transplanted from the Alps to the Apennines for this aim. We compared genetic variability and structure of native and introduced populations of *F. paralugubris* by AFLP assay and found that it was higher in the introduced populations, while native ones showed a higher diversity between nests. Overall, the genetic structure was dominated by among-worker variation regardless of different grouping arrangement (Alps vs. Apennine, native vs. introduced).

Abstract: The *Formica rufa* group comprises several ant species which are collectively referred to as "red wood ants" and play key roles in boreal forest ecosystems, where they are ecologically dominant and greatly influence habitat dynamics. Owing to their intense predatory activity, some of these species are used as biocontrol agents against several forest insect pests and for this aim in Italy, nearly 6000 ant nests were introduced from their native areas in the Alps to several Appeninic sites during the last century. In this work, we assessed and compared the genetic variability and structure of native and introduced populations of *F. paralugubris*, thus evaluating the extent of genetic drift that may have occurred since the time of introduction, using amplified fragment length polymorphism (AFLP) markers. PCR amplification with a fam_EcoRI-TAC/MseI-ATG primers combination produced a total of 147 scorable bands, with 17 identified as outlier loci. The genetic variation was higher in the introduced population compared to the native ones that, on the other hand, showed a higher diversity between nests. AMOVA results clearly pointed out that the overall genetic structure was dominated by among-worker variation, considering all populations, the Alpine vs. Apennine groups and the comparison among native and related introduced populations (all ranging between 77.84% and 79.84%). Genetic analyses unveiled the existence of six main different groups that do not entirely mirror their geographic subdivision, pointing towards a wide admixture between populations, but, at the same time, rapid diversification of some Apennine populations. Future studies based on high-throughput genomic methods are needed to obtain a thorough understanding of the effects of environmental pressure on the genetic structure and mating system of these populations.

Keywords: red wood ants; Foreste Casentinesi National Park; introduced species; AFLP; genetic diversity

1. Introduction

Red wood ants (RWA) are ecologically dominant species native of boreal forests of Central and Northern Europe [1]. They belong to the *Formica rufa* Palearctic complex, which in Western Europe comprises at least six species: *F. rufa* (Linneus, 1758), *F. aquilonia* (Yarrow, 1955), *F. lugubris* (Zetterstedt, 1838), *F. paralugubris* (Seifert, 1996), *F. polyctena* (Foerster,

1850) and *F. pratensis* (Retzius, 1783). All these species are characterised by the ability to build large aboveground nest mounds, and by the red and black coloration of their bodies. From the ecological point of view, RWA are keystone species, and they deeply impact the functioning of their forest ecosystems across multiple trophic levels [2]. These species influence the dynamics of arthropod communities through predation and competition [3], the structure of plant and lichen communities through their action on aphids, parasites and herbivores or propagule dispersion [4,5] and, ultimately, they affect nutrient cycling and soil functioning [6,7]. Despite the key role and abundance in most of their distribution range, the conservation status of this species is raising increasing concerns as there is evidence of local decline and even local extinction [8,9].

Owing to their intense predatory activity, some RWA species have been employed as biocontrol agents against several forest insect pests [10]. For this purpose, in Italy and Germany, nests of these species were transplanted from their original areas to other sites where they were formerly absent [11]. Between 1958 and 1972, more than 6000 nests of *F. lugubris/paralugubris*, *F. polyctena* and *F. aquilonia* were repeatedly transferred from the Alps to the Apennines and other Italian mountainous areas [12]. These introductions were carried out without considering the possible risks caused by the numerical reduction of the native populations, nor their possible negative impact on the newly occupied ecosystems, and it is worth mentioning that nowadays this practice is forbidden [13]. While in some cases the introductions resulted in viable populations that started to expand, actively preying upon the arthropod fauna in the newly occupied areas, other attempts failed [12,13]. The status of most of these introduced populations, similar to that of the native populations in the Alps, is unknown, and certainly calls for further studies [9]. Studying the ecology of these introduced populations is scientifically relevant, as they are a sort of a unique long-term ecological experiment that can provide important information on community dynamics, the effect of the introduction of dominant species, and population genetics.

Formica paralugubris was one of the most frequently introduced species in the Italian peninsula, but also to Canada [14,15] and was described for the first time as a sibling species of *F. lugubris* by Seifert [16]. Since its description, autochthonous populations of this species were discovered in the Pyrenees, Alps and the Jura mountains, at elevations ranging from 600 m to 2200 m asl [17]. *Formica paralugubris* can form huge supercolonies, composed of tens to hundreds of interconnected nests that may cover areas of over 0.5 km^2 [18,19], where they outcompete other ant species and affect other arthropod communities [20]. Each nest may contain hundreds of reproductive queens and two different reproductive strategies have been described. In one case, sexuals mate and remain within their natal colony, while in the other they perform mating flights that ensure long-distance dispersal [21]. The two strategies have a deep impact on the genetic structure of colonies, nest networks and relatedness within the nest. In general, long-distance dispersal through mating flights is rare, while budding is by far the most common colony reproduction mechanism. As a consequence, nearby nests in a supercolony are usually genetically closer than distant ones [22]

In the present study, we investigated the genetic variability in two native and four introduced populations (two from each native population) of *F. paralugubris*, using amplified fragment length polymorphism (AFLP) analysis. AFLPs are a PCR-based highly replicable dominant marker that allows rapid screening of genetic diversity and intraspecific variation without a priori sequence knowledge and at a low cost [23]. Our goal was to compare the genetic structure of native and introduced populations and evaluate the extent of genetic drift that may have occurred since the time of introduction 70 years ago. Furthermore, these local populations are reproductively isolated from those in the Alps and subjected to different selection pressures than those experienced by the native populations, given the lower latitude and altitude of their habitats. Based on these premises, we hypothesised that (i) the introduced populations show signatures of divergence from their native population while retaining some similarity among them; that (ii) the genetic variability is higher in the native Alpine populations than in those introduced in the Apennines. Our results will be a blueprint for future studies based on high-throughput genomic methods applied to selected nest samples; this will provide sufficient power for a thorough understanding of the effects of environmental pressure on the genetic structure, and mating system of these ant populations.

2. Materials and Method

2.1. Sampling Site

Formica paralugubris ants were collected from six populations—two native from the Alps and four introduced in the Apennine (Figure 1)—from June to August 2019. The Alpine populations were from Giovetto di Paline Nature Reserve (abbreviated as GP, 45°57′57″ N, 10°7′48″ E) and Baradello (abbreviated BA, 46°08′40″ N, 10°10′08″ E). We identified and sampled the populations located in the same areas reported as the origin of the introduced nests, and we assumed they are the descendants of the original ones. The introduced populations were from the Foreste Casentinesi, Monte Falterona and Campigna National Park, where nests of this species had been repeatedly introduced between 1958 and 1964. These populations were from the following locations: Avorniolo Alto (abbreviated as AA, 43°52′03″ N, 11°44′15″ E), Fosso Fresciaio (abbreviated FF, 43°51′23″ N, 11°44′36″ E), Le Cullacce (abbreviated LC, 43°51′42″ N, 11°45′50″ E) and La Lama (abbreviated LM, 43°49′36″ N, 11°48′24″ E). Populations from AA and FF originated from nests collected at GP, while the ones at LC and LM were from nests collected at BA. The Alpine sites were characterised by mixed forest, composed of a dominant conifer species, the Norway spruce (*Picea abies*, H.Karst., 1881), and beech (*Fagus sylvatica*, Linneaus, 1753). LC and FF were also characterised by mixed forest, but the local dominant conifer species is the silver fir (*Abies alba*, Miller, 1759), while AA and LM are covered by almost pure silver fir stands. More details on the ecology of the introduced populations and the history of their introduction can be found in [11,13].

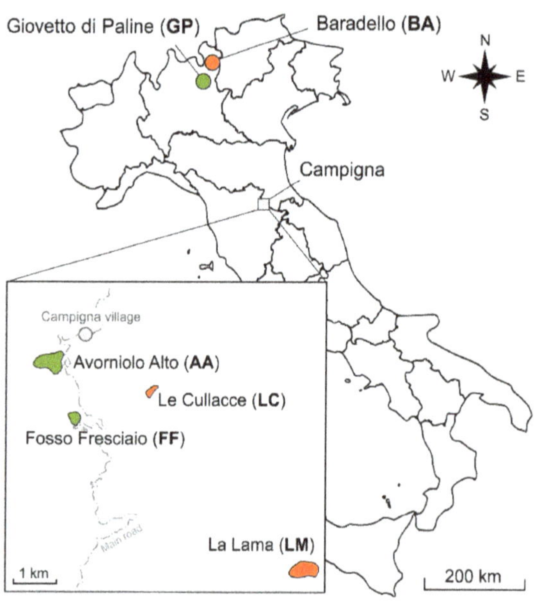

Figure 1. Geographic location of *F. paralugubris* populations investigated in this study.

For each population, we selected ten nests located at least 50 m apart, and ten ants were collected from the surface of each nest, for a total of 600 workers. The geographic position of the sampled nests was recorded by a GPS locator (Garmin eTrex® 10, accuracy ~3 m, Garmin, Olathe, KS, USA). Ants were stored in plastic test tubes (8 mL volume) filled with pure reagent-grade ethanol. All samples were transferred to the laboratory within 24 h from collection and stored at −80 °C for genetic analysis.

One individual per nest at BA and LM was identified as F. paralugubris following the molecular method developed by Bernasconi et al. [24], while species identity of ants collected in the other sites had been corroborated in previous studies [15,25].

2.2. *AFLP Fingerprinting*

Genomic DNA was extracted from a single adult thorax (10 individuals per nest), with a modified Chelex method [26]. Thorax without legs was ground into an Eppendorf plastic tube (1.5 μL) filled with 100 μL of 5% Chelex (BioRad, Hercules, CA, USA) solution (5 mg Chelex 100 resin in 50 mL ddH_2O). The sample was gently stirred and incubated at 56 °C for 4 h, adding 4 μL of Proteinase K. After centrifugation at 13,000 for 3 min, the upper aqueous supernatant was precipitated with 100% ethanol and 3 M sodium acetate, pH 5.2. This mixture was incubated at −20 °C for 20 min and then centrifuged at maximum speed for 10 min. The resulting pellet was washed by adding 70% ethanol and then centrifuged for 10 min at maximum speed. The supernatant was discarded and the pellet was airdried, resuspended in 30 μL of DNase-free water (Invitrogen, Waltham, MA, USA) and used as a template for PCR amplification. The extracted DNA was quantified using a Qubit 4 fluorometer (Invitrogen) with the dsDNA HS Assay kit (Invitrogen™ Q33231).

An analysis of AFLP divergence was performed following the procedure described by Coppi et al. [27]. The quality of AFLP profiles, and hence of the DNA extraction method, was preliminarily tested on 24 samples randomly selected from the 6 populations, evaluating six primer pair combinations. The fam_EcoRI-CTA/MseI-TTA combination was selected because of its highly comparable results across all the samples (in terms of good PCR products as well as of the number and size of the peaks obtained). Analysis of the AFLP profiles obtained by running capillary electrophoresis with the Applied Biosystems 3130xl platform was performed with GeneMarker v1.5 (SoftGenetics LLC, State College, PA, USA). A cut-off value, fixed at 5% of the maximum profile showed in the chromatograms, was determined after the analysis of replicate samples (reproducibility of the data was assessed by replicating 20 samples that were marked as duplicated and compared with the rest of the dataset in GeneMarker), considering only bands present in all the replicates. AFLP loci under selection (outliers) were screened using a Bayesian probability approach implemented in BayeScan v.2.01 [28]. The posterior probability of a given locus under selection was estimated, assuming that the locus frequencies within a population follow a multivariate β-distribution as a function of the multilocus fixation index value and of the average of locus frequencies of each locus between populations [29,30]. The analysis was set according to the software manual, considering 20 pilot runs with a length of 10,000 iterations each. The mean number of outlier loci was determined for each population.

2.3. *AFLP Analysis*

2.3.1. Genetic Diversity

Since AFLP are dominant markers, the presence or absence of every single fragment (100–2000 bp) was scored in each sample and coded by 1 or 0, creating a binary data matrix used to evaluate the within-population genetic variation as percentage of polymorphic loci (PL%), standard Nei's measure of genetic diversity ("h") as "average gene diversity over loci" [31] and Shannon's information index ("I") [32] for all populations in POPGENE v.1.32 [33].

The analysis of molecular variance (AMOVA) was performed using the ARLEQUIN v. 2.000 [34] to determine the partitioning of the overall genetic variation among all populations, between the Alpine and Apennine groups, and between the native populations and their related introduced ones, considering different hierarchical levels: between populations, among nests within populations and among workers. The analyses were performed separately, considering different hypothetical population groupings tested in terms of the variance components and the percentage of total expressed variation. Genetic diversity among nests within each population was assessed evaluating the fixation index ($F_{ST\text{-}POP}$), as difference in the allele frequency between nests.

2.3.2. Genetic Distance and Structure

A pairwise distances matrix among workers was computed following the Tamura–Nei method [35] and then a clustering analysis among nests, based on unweighted pair-group method with arithmetic means (UPGMA), was carried out in POPGENE and MEGA v.10.2.4 [36].

The analysis of population structures was performed following a model-based Bayesian clustering method implemented in STRUCTURE v.2.3.4 [37]. An admixture and shared allele frequency model was used to determine the number of clusters (K), assumed to be in the range between 2 and 12, with 10 replicate runs for each potential group. For each run, the initial burn-in period was set to 20,000 with 200,000 MCMC (Markov chain Monte Carlo) iterations, with no prior information on the origin of individuals. The best fit for the number of clusters, K, was determined using the Evanno method [38], as implemented in STRUCTURE HARVESTER [39]. STRUCTURE results were then elaborated using the R\pophelper package to align cluster assignments across replicate analyses and produce visual representations of cluster assignments.

3. Results

The AFLP dataset resulted from the analysis of 534 samples (Table 1), since the other 66 were lost (two whole nests at BA8 and LC10) because of bad quality amplification. The selected primers fam_EcoRI-TAC/MseI-ATG produced a total of 147 scorable bands with molecular weights ranging from approximately 40 to 550 base pairs. The BayeScan analysis identified 17 outlier loci that had a posterior probability higher than 0.8 (at a threshold of log10 PO ranging from 0.5 and 3), representing the 5% of all analysed loci. The highest locus diversity was found in LC, LM and AA with a mean number of outlier loci of 6.88, 6.34 and 6.26, respectively, whereas the lowest number was found at BA (3.21). GP and FF showed intermediate values amounting to 4.73 and 4.86, respectively. For the sake of clarity, no locus was excluded in any of the analyses.

Table 1. Genetic diversity traits among populations: effective number of nests (N_N) and of workers (N_W) analysed; number of polymorphic loci (NPL) percentage of polymorphic loci (PL%); Nei's genetic diversity (h, ±S.E.); fixation index ($F_{ST\text{-}POP}$ ± S.E) calculated among nests of the same population Shannon's information index (I, ±S.E.).

	Population	locality	N_N	N_w	NPL	PL(%)	$F_{ST\text{-}POP}$	h	I
GP	Native	Alps, Giovetto Paline	10	100	99	67.35	0.207 ± 0.003	0.104 ± 0.007	0.160 ± 0.020
AA	Transplanted	Apennine, Avorniolo	10	100	115	78.23	0.157 ± 0.006	0.184 ± 0.014	0.235 ± 0.027
FF	Transplanted	Apennine, Fosso Fresciaio	10	75	100	68.03	0.208 ± 0.005	0.104 ± 0.01	0.155 ± 0.023
BA	Native	Alps, Baradello	9	90	76	51.70	0.208 ± 0.002	0.102 ± 0.004	0.139 ± 0.021
LC	Transplanted	Apennine, Le Cullacce	9	87	133	90.48	0.147 ± 0.006	0.205 ± 0.011	0.231 ± 0.022
LM	Transplanted	Apennine, La Lama	10	82	118	80.71	0.183 ± 0.006	0.191 ± 0.012	0.199 ± 0.021

The percentage of polymorphic loci (PL%) among population ranged from a maximum of 90.48% (LC) to a minimum of 51.7% (BA). A summary of all the diversity measured is reported in Table 1. Genetic diversity (h) varied among populations and the highest h values were observed in the two introduced populations, AA and LC. If populations were divided according to a latitudinal gradient, the Apennine ones had higher values of diversity compared to their Alpine counterpart. A comparable trend can be also observed for either for PL%, and I values.

When we considered the population differentiation (measured by $F_{ST\text{-}POP}$, h and I), it appeared that the two native populations had the lowest values of genetic diversity (Table 1) but the highest among their nests. AMOVA showed significant differentiation among all populations and between latitudinal (Alpine vs. Apennine) groups (F_{ST} = 0.206, $p < 0.001$; F_{ST} = 0.221, $p < 0.001$, respectively). Nearly the same situation was found between the native populations and their related introduced ones ($F_{ST} = 0.217$, $p < 0.001$, $F_{ST} = 0.218$, $p < 0.001$, respectively, for BA vs. LC + LM and GP vs. AA + FF). A genetic structure dominated by among-worker variation (see Table 2) was evident at all levels and did not correspond to comparable among-nets variations.

Table 2. Partitioning of genetic variation. AMOVA was performed, testing a four groupings scenario to test the differentiation between 534 individual samples from 6 populations. The table shows: degrees of freedom (df), sum of squared deviations, estimated variance components, percentages of total variance contributed by each component, three different fixation indexes and the probability of obtaining a more extreme component estimate by chance alone (P).

Source of Variation		df	Sum of Squares	Variance Components	% Variation	Fix.Index	*p*-Values
All population							
	Among populations	5	728.16	1.56	12.15	$F_{CT} = 0.121$	<0.001
	Among nests within populations	43	878.42	1.02	8.05	$F_{SC} = 0.091$	<0.001
	Among workers	435	4465.26	10.26	79.84	$F_{ST} = 0.201$	<0.001
	Total	483	6071.87	12.85	100		
Alpine vs. Apennine							
	Among groups	1	294.21	1.13	8.6	$F_{CT} = 0.085$	<0.05
	Among nests within groups	47	1312.33	1.78	13.56	$F_{SC} = 0.148$	<0.05
	Among workers	435	4465.22	10.26	77.84	$F_{ST} = 0.221$	<0.001
	Total	483	6071.81	13.17	100		
BA vs. (LC + LM)							
	Among groups	1	182.21	1.23	9.23	$F_{CT} = 0.092$	<0.001
	Among nests within groups	28	714.33	1.68	12.57	$F_{SC} = 0.138$	<0.001
	Among workers	239	2504.91	10.48	78.20	$F_{ST} = 0.217$	<0.001
	Total	268	3401.45		100		
GP vs. (AA + FF)							
	Among groups	1	144.39	0.94	7.65	$F_{CT} = 0.076$	<0.001
	Among nests within groups	28	701.93	1.75	14.20	$F_{SC} = 0.153$	<0.001
	Among workers	235	2269.92	9.65	78.15	$F_{ST} = 0.218$	<0.001
	Total	264	3116.24	12.35	100		

The UPGMA-based dendrogram (Figure 2) showed that the six populations clustered into two major groups: one comprising the native Alpine populations and FF, and one including the other Apenninic populations (LC, LM and AA). When looking at the second cluster, the three Apennine populations clustered together, with LC and LM very close to each other.

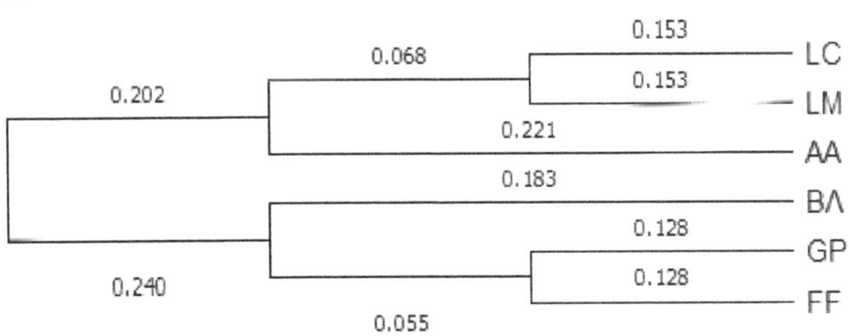

Figure 2. UPGMA dendrogram of investigated populations based. Studied populations: LC, Le Cullacce; LM, La Lama; AA, Avorniolo Alto; BA, Baradello; GP, Giovetto di Paline; FF, Fosso Fresciaio.

The population genetic kinship obtained with the UPGMA-based tree was subsequently confirmed by STRUCTURE results (Figure 3). Worker genotypes were assigned to a cluster with a probability >0.6. The optimum number of populations, K, estimated (Figure S1) according to the Evanno method, was six. This suggests the occurrence of six populations (i.e., gene pools) which were nonetheless clearly admixed and do not correspond to the geographic origin of the samples. The two Alpine populations (GP and BA) were highly similar and quite different from the others. FF population was relatively close to the Alpine ones, due to the sharing of several alleles, especially with BA, but also showed some distinctiveness compared to the other Apennine populations. AA and LC seemed quite differentiated from their population of origin and similar to each other, while LM showed all the six gene pools highly admixed.

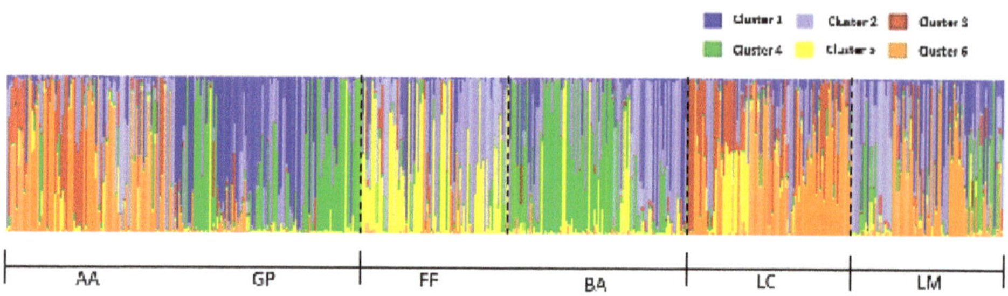

Figure 3. STRUCTURE plot of the 534 worker genotypes. Each individual is represented by a vertical line; the genetic clusters identified are numbered from 1 to 6 and marked with different colours. Population affiliation is also indicated: LC, Le Cullacce; LM, La Lama; AA, Avorniolo Alto; BA, Baradello; GP, Giovetto di Paline; FF, Fosso Fresciaio.

4. Discussion and Conclusions

This study assessed the variation of genetic diversity and structure across *F. paralugubris* populations translocated to the Apennines and as compared with their Alpine populations of origin 70 years after the introduction. We successfully used an AFLP assay for detecting overall differentiation among and within populations, on the basis of 147 loci.

About 5% of the identified loci can be considered outliers and although it was impossible to directly associate them specifically with any of the surveyed populations, it is important to note that, overall, they were more numerous in the introduced ones. The different pedoclimatic conditions, altitude, average temperature and vegetation between Alps and Apennines may account for the observed difference in outlier frequency and expression [40], although this finding certainly requires further study.

Our results also confirmed the existence of considerable genetic admixture in all the surveyed populations, which was expected, based on the observations on the reproductive behaviour reported by Frizzi et al. [41] and the findings on other European *F. paralugubris* populations as well as other RWA species [19]. The low genetic variance among nests (8% of total) suggests gene flow between them and is also consistent with budding reproductive behaviour, characterised by mating occurring in close proximity or even within the natal nest [20]. On the contrary, the high intra-nest genetic variance, which amounts to nearly 80% of the total observed variation, confirms that multiple unrelated or weakly unrelated queens may inhabit the same nest [22]. Finally, the low among-populations variability (12% of total variance) can be explained by considering that all the colonies belong to the same native macro area in the Alps, and have been randomly separated nearly seven decades ago, which is a short time in terms of natural selection [42]. If we compare the genetic variance between Apennine and Alpine populations, we see nearly the same trend, with the major proportion of variance being partitioned among workers within each population (78%). However, we have to point out that, in this respect, the Alpine populations were under-represented, since only 20 nests were analysed, as opposed to the 40 from the Apennines. Another important point to consider is that we cannot be completely sure that the colonies we sampled as native populations are really the direct descendants of the ants transplanted 70 years ago. We sampled exactly the same areas, but we could not exclude possible supercolony replacement events during the last decades.

Contrary to our expectation, the introduced LM, LC, and AA had the greatest genetic diversity, while the lowest values were detected in the native Alpine populations. These were characterised by a low number of polymorphic loci and the highest values of $F_{ST\text{-}POP}$

genetic distance among nests. This finding suggests that the gene pool of these populations might have undergone a reduction through the years, presumably as a consequence of demographic decline [43]. This point deserves careful attention and cannot be overlooked. Paradoxically, while the populations introduced to the Apennines are expanding [13], the status of the populations in the Alps is poorly known, and there are no indications of whether they are stable, increasing, or decreasing [9]. It is likely that the massive collection of nests carried out during the last century for transplantation [12] could have caused a non-negligible reduction of their size and, hence, genetic diversity [44]. However, it should be kept in mind that we have no information on the genetic structure and variability of the original Alpine populations at the time of nest collection, making further speculations difficult.

As for the populations in the Apennines, there is no doubt that the new habitat differed in many ways from that in the original locations in the Alps, both as composition of the forest stands (i.e., presence of silver fir instead of Norway spruce) and, more importantly, in climate, being located more than 250 km south of the Alps. The concurrent action of these factors could have imposed local selective pressures underlying the observed changes. Moreover, we also know that most of the original nest mounds introduced (20 at AA, 19 at FF, 27 and 20 at LM and LC, respectively) had disappeared a few years later, and in nearly all areas the populations had reduced to three to six mounds [45]. We do not know exactly whether this was due to colony death, followed by selection and bottleneck effect, or it resulted from their fusion into few larger ones. The initial drastic reduction in nests number may have affected the genetic pool of the new populations in two opposite ways. If the reduction was due to nest aggregation, the initial genetic diversity should have been retained, at least initially. Alternatively, if the reduction was due to colony death, genetic drift would have occurred. The high genetic variability among Apennine populations detected by our analyses suggests an initial aggregation dynamic. However, further studies based on different genetic markers (e.g., single nucleotide polymorphisms, SNPs) able to infer colony structure, mating system and overall diversity are needed to fully address all these issues [46].

The population genetic makeup obtained with STRUCTURE suggests the existence of six strongly admixed groups that do not correspond to the geographic origin. Admixture included elements from both the Apennine and the Alps, though AA, LC, LM and FF were seemingly different from their native ones. One point that needs to be carefully considered, however, is that the populations introduced to the Apennines 70 years ago may represent an important source of genetic diversity. Considering the low genetic variability of the native populations and increasing evidence of local declines and extinction in several parts of the species range, Alps included, these findings may have relevant implications for the conservation of RWA. Further studies are needed to better assess the status of the Alpine populations of *F. paralugubris*, while transplanted populations should be preserved through specific conservation policies and plans.

Finally, AFLP analysis turned out to be an efficient and cost-effective tool for assessing genetic diversity and variation in these populations, at least for a first screening based on population genetic variance. Compared to other markers such as RAPD (random amplified polymorphic DNA) and SSR (short sequence repeat) based on a polymerase chain reaction (PCR) step, AFLP can rapidly generate a great number of polymorphisms with a high number of assayed loci in virtually any organism, being often used for diversity and population genetics studies in ants and other insects [47–49]. The recent sequencing of the first RWA genome by Nouhaud et al. [50] opened up the prospect of using multiple next generation sequencing (NGS) approaches based on SNPs detection for population and phylogenetic studies of this ant group. For example, Portinha et al. [51] performed whole-genome resequencing of several workers to infer divergence histories among heterospecific populations of *F. polyctena* and *F. aquilonia*. As far as our populations are concerned, a double-digest restriction-site-associated DNA sequencing (ddRADseq) approach could be a powerful tool to compare the genetic variation, structure and mating system across populations [52,53] along with genome-wide association studies (GWAS) focused on the identification of the genes that are actively selected by the environmental factors and integrating phenotypic information [54].

In conclusion, this study provides the first information about variation in genetic diversity of native and introduced populations of *F. paralugubris* in Italy. The results may not only assist the conservation of native RWA populations but also better our understanding of the dynamics experienced by ant species when transplanted outside their distribution range.

Supplementary Materials: The following supporting information can be downloaded at: https://www.mdpi.com/article/10.3390/ani12223165/s1, Figure S1: Plot of Delta-K values according to Evanno method for clusters identification.

Author Contributions: Conceptualization, A.C., M.Z., R.F. and G.S.; Methodology, P.B. and A.C.; Software, A.C. and F.F.; Formal analysis, A.M., F.F. and G.S.; Data curation, A.M. and P.B.; Writing – original draft, A.M.; Supervision, A.C. and R.F.; Project administration, G.S. All authors have read and agreed to the published version of the manuscript.

Funding: This research received no external funding.

Institutional Review Board Statement: Not applicable.

Informed Consent Statement: Not applicable.

Data Availability Statement: Data will be made available on request.

Conflicts of Interest: The authors declare no conflict of interest.

References

1. Stockan, J.; Robinson, E.; Trager, J.; Yao, I.; Seifert, B. Introducing wood ants: Evolution, phylogeny, identification and distribution. In *Wood Ant Ecology and Conservation (Ecology, Biodiversity and Conservation)*, 1st ed.; Stockan, J., Robinson, E., Eds.; Cambridge University Press: Cambridge, UK, 2016; pp. 1–36. [CrossRef]
2. Risch, A.C.; Ellis, S.; Wiswell, H. Where and why? Wood ant population ecology. In *Wood Ant Ecology and Conservation (Ecology, Biodiversity and Conservation)*, 1st ed.; Stockan, J., Robinson, E., Eds.; Cambridge University Press: Cambridge, UK, 2016; pp. 81–105. [CrossRef]
3. Punttila, P.; Niemelä, P.; Karhu, K. The impact of wood ants (Hymenoptera: Formicidae) on the structure of invertebrate community on mountain birch (*Betula pubescens* ssp. *czerepanovii*). *Ann. Zool. Fenn.* **2004**, *41*, 429–446.
4. Ohashi, M.; Kilpeläinen, J.; Finér, L.; Risch, A.C.; Domisch, T.; Neuvonen, S.; Niemelä, P. The effect of red wood ant (*Formica rufa* group) mounds on root biomass, density, and nutrient concentrations in boreal managed forests. *Eur. J. For. Res.* **2007**, *12*, 113–119. [CrossRef]
5. Di Nuzzo, L.; Masoni, A.; Frizzi, F.; Bianchi, E.; Castellani, M.B.; Balzani, P.; Benesperi, R. Red wood ants shape epiphytic lichen assemblages in montane silver fir forests. *iForest* **2022**, *15*, 71–76. [CrossRef]
6. Frouz, J.; Jílková, V.; Sorvari, J. Contribution of wood ants to nutrient cycling and ecosystem function. In *Wood Ant Ecology and Conservation (Ecology, Biodiversity and Conservation)*, 1st ed.; Stockan, J., Robinson, E., Eds.; Cambridge University Press: Cambridge, UK, 2016; pp. 81–105. [CrossRef]
7. Balzani, P.; Masoni, A.; Venturi, S.; Frizzi, F.; Bambi, M.; Fani, R.; Santini, G. CO_2 biogeochemical investigation and microbial characterization of red wood ant mounds in a Southern Europe montane forest. *Soil Biol. Biochem.* **2022**, *166*, 108536. [CrossRef]
8. Mabelis, A.A.; Korczyńska, J. Long-term impact of agriculture on the survival of wood ants of the *Formica rufa* group (Formicidae). *J. Insect Conserv.* **2016**, *20*, 621–628. [CrossRef]
9. Balzani, P.; Dekoninck, W.; Feldhaar, H.; Freitag, A.; Frizzi, F.; Frouz, J.; Santini, G. Challenges and a call to action for protecting European red wood ants. *Cons. Biol.* **2022**, e13959. [CrossRef]
10. Gosswald, K. *Die Rote Waldameise im Dienste der Waldhygiene: Forstwirtschaftliche Bedeutung, Nutzung, Lebensweise, Zucht, Wermehrung und Schutt*; Metta Kinau Verlag: Lüneburg, Germany, 1951; p. 160.
11. Ronchetti, G.; Mazzoldi, P.; Groppali, R. *Venticinque Anni di Osservazioni Sui Trapianti di Formica Lugubris Zett. Dalla Alpi Alle Foreste Demaniali Casentinesi (Italia centrale): (Hymen. Formicidae)*; Università di Pavia: Pavia, Italy, 1986; p. 121.
12. Ronchetti, G.; Groppali, R. *Quarantacinque Anni di Protezione Forestale con Formica lugubris Zett. (Hymenoptera Formicidae). L'esperienza di Monte d'Alpe (Appennino Ligure in Provincia di Pavia)*; Istituto di Entomologia dell'Università di Pavia: Pavia, Italy, 1995; p. 271.
13. Frizzi, F.; Masoni, A.; Quilghini, G.; Ciampelli, P.; Santini, G. Chronicle of an impact foretold: The fate and effect of the introduced *Formica paralugubris* ant. *Biol. Inv.* **2018**, *20*, 3575–3589. [CrossRef]
14. Seifert, B. The supercolonial European wood ant *Formica paralugubris* Seifert, 1996 (Hymenoptera: Formicidae) introduced to Canada and its predicted role in Nearctic forests. *Myrmecol. News* **2016**, *22*, 11–20.

15. Masoni, A.; Frizzi, F.; Natali, C.; Bernasconi, C.; Ciofi, C.; Santini, G. Molecular identification of imported red wood ant populations in the Campigna biogenetic nature Reserve (Foreste Casentinesi national Park, Italy). *Cons. Gen. Res.* **2019**, *11*, 231–236. [CrossRef]
16. Seifert, B. *Formica paralugubris* nov. spec.—A sympatric sibling species of *Formica lugubris* from the western Alps (Insecta: Hymenoptera: Formicoidea: Formicidae). *Reichenbachia* **1996**, *35*, 193–201.
17. Bernasconi, C.; Maeder, A.; Freitag, A.; Cherix, D. *Formica paralugubris* (Hymenoptera, Formicidae) in the Italian Alps from new data and old data revisited. *Myrmecol. News* **2006**, *8*, 251–256.
18. Cherix, D. Note preliminaire sur la structure, la phenologie et le regime alimentaire d'une super-colonie de *Formica lugubris* Zett. *Insect. Soc.* **1980**, *27*, 226–236. [CrossRef]
19. Chapuisat, M.; Goudet, J.; Keller, L. Microsatellites reveal high population viscosity and limited dispersal in the ant *Formica paralugubris*. *Evolution* **1997**, *51*, 475–482. [CrossRef] [PubMed]
20. Holzer, B.; Meunier, J.; Keller, L.; Chapuisat, M. Stay or drift? Queen acceptance in the ant *Formica paralugubris*. *Insect. Soc.* **2008**, *55*, 392–396. [CrossRef]
21. Cherix, D. Red wood ants. *Ethol. Ecol. Evol.* **1991**, *3*, 165. [CrossRef]
22. Chapuisat, M.; Keller, L. Extended family structure in the ant *Formica paralugubris*: The role of the breeding system. *Behav. Ecol. Soc.* **1999**, *46*, 405–412. [CrossRef]
23. Blears, M.J.; De Grandis, S.A.; Lee, H.; Trevors, J.T. Amplified fragment length polymorphism (AFLP): A review of the procedure and its applications. *JIMB* **1998**, *21*, 99–114. [CrossRef]
24. Bernasconi, C.; Pamilio, P.; Cherix, D. Molecular markers allow sibling species identification in red wood ants (*Formica rufa* group). *Syst. Entomol.* **2010**, *35*, 243–249. [CrossRef]
25. Balzani, P.; Vizzini, S.; Frizzi, F.; Masoni, A.; Lessard, J.P.; Bernasconi, C.; Santini, G. Plasticity in the trophic niche of an invasive ant explains establishment success and long-term coexistence. *Oikos* **2021**, *130*, 691–696. [CrossRef]
26. Lienhard, A.; Schäffer, S. Extracting the invisible: Obtaining high quality DNA is a challenging task in small arthropods. *PeerJ* **2019**, *7*, e6753. [CrossRef]
27. Coppi, A.; Cecchi, L.; Mengoni, A.; Phustahija, F.; Tomović, G.; Selvi, F. Low genetic diversity and contrasting patterns of differentiation in the two monotypic genera *Halacsya* and *Paramoltkia* (Boraginaceae) endemic to the Balkan serpentines. *Flora* **2014**, *209*, 5–14. [CrossRef]
28. Foll, M.; Fischer, M.C.; Heckel, G.; Excoffier, L. Estimating population structure from AFLP amplification intensity. *Mol. Ecol.* **2010**, *19*, 4638–4647. [CrossRef] [PubMed]
29. Burr, T.L. Quasi-equilibrium theory for the distribution of rare alleles in a subdivided population: Justification and implications. *Theor. Popul. Biol.* **2000**, *57*, 297–306. [CrossRef] [PubMed]
30. Yang, A.H.; Dick, C.W.; Yao, X.H.; Huang, H.W. Impacts of biogeographic history and marginal population genetics on species range limits: A case study of *Liriodendron chinense*. *Sci. Rep.* **2016**, *6*, 25632. [CrossRef]
31. Nei, M. *Molecular Evolutionary Genetics*; Columbia University Press: New York, NY, USA, 1987; p. 512.
32. Lewontin, R.C. Evolutionary biology: The apportionment of human diversity. In *Evolutionary Biology*; Dobzhansky, T., Hecht, M.K., Steere, W.C., Eds.; Springer: New York, NY, USA, 1995; pp. 381–398. [CrossRef]
33. Yeh, F.; Yang, R.; Boyle, T.; Ye, Z.; Mao, J. *POPGEN Ver. 1.32. The User-Friendly Software for Population Genetic Analysis*; University of Alberta, Molecular Biology and Biotechnology Center: Edmonton, AB, Canada, 1997.
34. Schneider, S.; Roessli, D.; Excer, L. *Arlequin: A Software for Population Genetics Data Analysis*; University of Geneva, Genetics and Biometry Laboratory: Geneva, Switzerland, 2000.
35. Tamura, K.; Nei, M. Estimation of the number of nucleotide substitutions in the control region of mitochondrial DNA in humans and chimpanzees. *Mol. Biol. Evol.* **1993**, *10*, 512–526. [CrossRef]
36. Kumar, S.; Nei, M.; Dudley, J.; Tamura, K. MEGA: A biologist-centric software for evolutionary analysis of DNA and protein sequences. *Brief. Bioinform.* **2008**, *9*, 299–306. [CrossRef]
37. Hubisz, M.J.; Falush, D.; Stephens, M.; Pritchard, J.K. Inferring weak population structure with the assistance of sample group information. *Mol. Ecol. Res.* **2009**, *9*, 1322–1332. [CrossRef]
38. Evanno, G.; Regnaut, S.; Goudet, J. Detecting the number of clusters of individuals using the software STRUCTURE: A simulation study. *Mol. Ecol.* **2005**, *14*, 2611–2620. [CrossRef]
39. Earl, D.A.; Von Holdt, B.M. STRUCTURE HARVESTER: A website and program for visualizing STRUCTURE output and implementing the Evanno method. *Conserv. Genet. Resour.* **2012**, *4*, 359–361. [CrossRef]
40. Menchetti, M.; Talavera, G.; Cini, A.; Salvati, V.; Dincă, V.; Platania, L.; Dapporto, L. Two ways to be endemic. Alps and Apennines are different functional refugia during climatic cycles. *Mol. Ecol.* **2021**, *30*, 1297–1310. [CrossRef]
41. Frizzi, F.; Masoni, A.; Santedicola, M.; Servini, M.; Simoncini, N.; Palmieri, J.; Santini, G. Intraspecific relationships and nest mound shape are affected by habitat features in introduced populations of the red wood ant *Formica paralugubris*. *Insects* **2022**, *13*, 198–204. [CrossRef] [PubMed]
42. Kulmuni, J.; Nouhaud, P.; Pluckrose, L.; Satokangas, I.; Dhaygude, K.; Butlin, R.K. Instability of natural selection at candidate barrier loci underlying speciation in wood ants. *Mol. Ecol.* **2020**, *29*, 3988–3999. [CrossRef] [PubMed]
43. Maky-Petays, H.; Zakharov, A.; Viljakainen, L.; Corander, J.; Pamilo, P. Genetic changes associated to declining populations of Formica ants in fragmented forest landscape. *Mol. Ecol.* **2005**, *14*, 733–742. [CrossRef] [PubMed]

44. Allendorf, F.W.; England, P.R.; Luikart, G.; Ritchie, P.A.; Ryman, N. Genetic effects of harvest on wild animal populations. *Trends Ecol. Evol.* **2008**, *23*, 327–337. [CrossRef]
45. Groppali, R.; Crudele, G. Le formiche del gruppo *Formica rufa* trapiantate nel Parco nazionale delle Foreste Casentinesi, Monte Falterona e Campigna. *Quad. Stud. Nat. Romagna* **2005**, *20*, 63–73.
46. Flanagan, S.P.; Jones, A.G. The future of parentage analysis: From microsatellites to SNPs and beyond. *Mol. Ecol.* **2019**, *28*, 544–567. [CrossRef]
47. Gzyl, A.; Augustynowicz, E.; Mosiej, E.; Zawadka, M.; Gniadek, G.; Nowaczek, A.; Slusarczyk, J. Amplified fragment length polymorphism (AFLP) versus randomly amplified polymorphic DNA (RAPD) as new tools for inter-and intra-species differentiation within *Bordetella*. *J. Med. Microbiol.* **2005**, *54*, 333–346. [CrossRef]
48. Kulmuni, J.; Seifert, B.; Pamilo, P. Segregation distortion causes large-scale differences between male and female genomes in hybrid ants. *Proc. Nas. Acad. Sci. USA* **2010**, *107*, 7371–7376. [CrossRef]
49. Schluns, E.A.; Neumann, P.; Schluns, H.; Hepburn, H.R.; Moritz, R.F.A. Nestmate recognition and genetic variability among individuals from nests of the queenless ponerine ant, *Streblognathus aethiopicus* Smith (Hymenoptera: Formicidae). *Afr. Entomol.* **2006**, *14*, 95–102.
50. Nouhaud, P.; Beresford, J.; Kulmuni, J. Assembly of a hybrid *Formica aquilonia* × *F. polyctena* ant genome from a haploid male. *J. Hered.* **2022**, *9*, 353–359. [CrossRef]
51. Portinha, B.; Avril, A.; Bernasconi, C.; Helanterä, H.; Monaghan, J.; Seifert, B.; Nouhaud, P. Whole-genome analysis of multiple wood ant population pairs supports similar speciation histories, but different degrees of gene flow, across their European ranges. *Mol. Ecol.* **2022**, *31*, 3416–3431. [CrossRef] [PubMed]
52. Smith, C.C.; Weber, J.N.; Mikheyev, A.S.; Roces, F.; Bollazzi, M.; Kellner, K.; Mueller, U.G. Landscape genomics of an obligate mutualism: Concordant and discordant population structures between the leafcutter ant *Atta texana* and its two main fungal symbiont types. *Mol. Ecol.* **2019**, *28*, 2831–2845. [CrossRef] [PubMed]
53. Zhang, Y.M.; Vitone, T.R.; Storer, C.G.; Payton, A.C.; Dunn, R.R.; Hulcr, J.; Lucky, A. From pavement to population genomics: Characterizing a long-established non-native ant in North America through citizen science and ddRADseq. *Front. Ecol. Evol.* **2019**, *7*, 453. [CrossRef]
54. Jay, P.; Leroy, M.; Le Poul, Y.; Whibley, A.; Arias, M.; Chouteau, M.; Joron, M. Association mapping of colour variation in a butterfly provides evidence that a supergene locks together a cluster of adaptive loci. *Philos. Trans. R. Soc. B* **2022**, *377*, 20210193. [CrossRef]

animals

MDPI

Communication

"Guess Who's Coming to Dinner": Molecular Tools to Reconstruct *multilocus* Genetic Profiles from Wild Canid Consumption Remains

Edoardo Velli [1,*,†], Federica Mattucci [1,†], Lorenzo Lazzeri [2], Elena Fabbri [1], Giada Pacini [2], Irene Belardi [2], Nadia Mucci [1] and Romolo Caniglia [1]

1 Unit for Conservation Genetics (BIO-CGE), Italian Institute for Environmental Protection and Research (ISPRA), Via Cà Fornacetta 9, 40064 Ozzano dell'Emilia, Italy

2 Research Unit of Behavioural Ecology, Ethology and Wildlife Management, Department of Life Sciences, University of Siena, Via P.A. Mattioli 4, 53100 Siena, Italy

* Correspondence: edoardo.velli@isprambiente.it

† These authors equally contributed to the manuscript.

Simple Summary: Wolves and European wildcats are two iconic predator species that can live in overlapping ecological contexts and also share their habitats with their domestic free-ranging relatives, increasing the risk of anthropogenic hybridisation and its possible deleterious consequences. By applying a multidisciplinary approach, we morphologically and molecularly analysed the cat remains found in a canid faecal sample collected in a forested area of central Italy. Individual *multilocus* genotypes of both predator and prey were identified turning out to be, respectively, a wolf showing traces of dog ancestry at autosomal microsatellite *loci* and a domestic cat.

Abstract: Non-invasive genetic sampling is a practical tool to monitor pivotal ecological parameters and population dynamic patterns of endangered species. It can be particularly suitable when applied to elusive carnivores such as the Apennine wolf (*Canis lupus italicus*) and the European wildcat (*Felis silvestris silvestris*), which can live in overlapping ecological contexts and sometimes share their habitats with their domestic free-ranging relatives, increasing the risk of anthropogenic hybridisation. In this case study, we exploited all the ecological and genetic information contained in a single biological canid faecal sample, collected in a forested area of central Italy, to detect any sign of trophic interactions between wolves and European wildcats or their domestic counterparts. Firstly, the faecal finding was morphologically examined, showing the presence of felid hair and claw fragment remains. Subsequently, total genomic DNA contained in the hair and claw samples was extracted and genotyped, through a multiple-tube approach, at canid and felid diagnostic panels of microsatellite *loci*. Finally, the obtained individual *multilocus* genotypes were analysed with reference wild and domestic canid and felid populations to assess their correct taxonomic status using Bayesian clustering procedures. Assignment analyses classified the genotype obtained from the endothelial cells present on the hair sample as a wolf with slight signals of dog ancestry, showing a $qi = 0.954$ (C.I. 0.780–1.000) to the wolf cluster, and the genotype obtained from the claw as a domestic cat, showing a $qi = 0.996$ (95% C.I. = 0.982–1.000) to the domestic cat cluster. Our results clearly show how a non-invasive multidisciplinary approach allows the cost-effective identification of both prey and predator genetic profiles and their taxonomic status, contributing to the improvement of our knowledge about feeding habits, predatory dynamics, and anthropogenic hybridisation risk in threatened species.

Keywords: anthropogenic hybridisation; canid consumption; domestic cat; European wildcat; food habits; *multilocus* genotypes; non-invasive genetic sampling; predation; wolf

1. Introduction

Trophic relationships within ecosystems are key parameters in wildlife ecological studies [1]. Feeding habits, behaviour and dynamics of animal species, mainly assessed through diet analyses, can provide useful insights into intra- and inter-specific niche specialisation [2]. Moreover, such approaches are particularly suitable for a better comprehension of apex predator ecology, which plays a pivotal role in ecosystem equilibria [3].

During the last decades, diet analyses of vertebrate predators have been extensively conducted through multidisciplinary approaches, mainly based on morphological and molecular identifications of their prey remains contained in non-invasively collected faecal materials [4,5]. In particular, most molecular studies on diet have been focused on a simple taxonomic identification of the consumed species [6,7] using vertebrate broad range markers, such as the cytochrome b and the 16S subunit of the mitochondrial DNA, or applying metabarcoding techniques, to genotype prey DNA contained in predator scats [7–9]. However, the recent spread of anthropogenic hybridisation, which originated from the crossing between native and alien species or wild and domestic populations of the same species, requires the application of more powerful molecular tools. The use of panels of highly informative specific markers, such as microsatellite (STR) or single nucleotide polymorphism (SNP) *loci*, permits reconstructing the *multilocus* genetic profiles of both predator and prey when analysing faecal DNA for diet studies. Later on, unknown *multilocus* genotypes can be assigned to well-representative parental populations through statistical procedures to assess species, native populations or signs of admixture. Such data can also provide key information on differences in food dynamics, hunting behaviours and ecological relationships among taxa or between pure and admixed individuals of the same species [10]. This would be particularly useful, especially for two iconic and notoriously elusive Italian carnivores, the Apennine wolf (*Canis lupus italicus*) and the European wildcat (*Felis silvestris silvestris*), which often share their natural habitats and individual territories across the Italian Peninsula, and that can successfully mate with their domestic counterparts, the domestic dog (*C. l. familiaris*) and the domestic cat (*Felis catus*) [11,12].

The wolf and the European wildcat experienced very similar demographic scenarios in Italy, with protracted isolation south of the Alps and recurrent bottlenecks that made them sharply genetically differentiated from any other wolf or wildcat population [13,14]. Nowadays, both species are geographically re-expanding and numerically increasing trough the Peninsula, thanks to legal protection and their ecological plasticity [15,16], but they are still threatened by habitat fragmentation [16,17], accidental or illegal killings [18,19] and by anthropogenic hybridisation [20,21].

Even though distribution ranges and anthropogenic hybridisation, rates in Italy are continuously studied in wolves and European wildcats, especially through non-invasive genetic projects [16,17]. Interspecific relationships between the two species are poorly known, with only a few wolf predations recorded on felids [22,23]. Likewise, differences in behaviour, feeding strategies and diet composition between pure and admixed individuals in both carnivores are still scarcely studied, with only a few available data on a local scale [10].

In this case study, we analysed the DNA contained in the remains of a canid faecal deposition collected in a forested area of central Italy to determine the individual *multilocus* genetic profiles of both the predator and the prey. In particular, we exploited the availability of reliable forensic genetic protocols [24], well-performing panels of canid [25] and felid [26] unlinked autosomal STRs and robust statistical procedures [21] to genotype non-invasive samples, assess their origin and clarify if they had wild, domestic or admixed ancestry.

2. Materials and Methods

On 9 February 2020, we collected a canid faecal sample containing the consumption remains of a cat, consisting of hairs and claw fragments, in a wooded area of Tuscany in central Italy. The sample was initially preserved at −20 °C, subsequently stored in an oven at 80 °C for four hours to deactivate possible pathogens, and then frozen again at −20 °C until downstream morphological and molecular analyses.

After separating hairs from claw fragments with water and a 1 mm sieve, a preliminary morphological identification of the hair sample was conducted by observing cuticle and medulla patterns with a 100–400× zoom microscope [27].

Additionally, since stray dogs and cats, as well as wolf-dog and wildcat-domestic cat hybrids, were recorded in the study area, we stored a tuft of hairs ($n > 10$) and the claw fragment, respectively, in a paper envelope to search for possible canid intestinal endothelial cells and in 40 mL of 96% ethanol to be genetically analysed, thus clarifying the taxonomic status of both predator and prey. Total DNA contained in the hair or on its surface (likely containing endothelial intestinal cells of the predator) and in the claw samples was individually extracted using the Blood & Tissue Kit® (Qiagen), following the manufacturer's instructions.

Each DNA sample was amplified by Polymerase Chain Reaction (PCR) and genotyped through a multiple-tube approach [28] at diagnostic wolf and cat molecular markers. In particular, the diagnostic wolf marker panel included the following: (a) 39 unlinked autosomal microsatellite *loci* (STRs), discriminating among wolves, dogs, and their first three generations of hybrids [12,17]; (b) the *Amelogenin* marker, to molecularly sex the extracted DNA; (c) 4 Y-chromosome STRs (MS34A, MS34B, MSY41A and MS41B Sundqvist et al. 2001) determining the paternal haplotype in male individuals; and (d) a dominant 3-bp deletion at the *β-defensin* CBD103 gene (the *K-locus*) associated with black coat colour. The diagnostic cat marker panel included 29 domestic cat-derived dinucleotide STRs, discriminating among European wildcats, domestic cats, and their first two generations of hybrids [26,29,30].

All PCR reactions were performed in a total volume of 10 µL containing the following: 2 µL of DNA, 5 µL of MasterMix (Qiagen Multiplex Kit), 3 µL of Q-solution (Qiagen Multiplex Kit), 0.15–0.30 µL of primers adjusted to the volume with RNAse-free water. PCR products were analysed in an ABI 3130XL automated sequencer and allele sizes were estimated using Genemapper v.5.0 (Applied Biosystems).

Hair and claw DNA samples were extracted, amplified, and genotyped in three separate rooms dedicated to low-template DNA samples under sterile UV laminar flood hoods. Negative (no biological sample during extraction and no DNA in PCR) and positive (a wolf and a cat DNA sample of good quality and with known genotype) controls were included in each step to check for possible contaminations and correct allelic weights, respectively.

Consensus genotypes, amplification success (AS) and error (allelic dropout, ADO, false alleles, FA) rates were assessed from the four replicates per *locus* performed during the multiple-tube approach, using Gimlet v.1.3.3 [31].

Genotype reliability was calculated by RelioType [32], considering an acceptance value of $R \geq 95\%$.

Reliable genotypes were then assigned to their populations of origin (wolf or dog; European wildcat or domestic cat) through a first explorative Principal Coordinate Analysis (PCoA) using GenAlEx v.6 [33] and Bayesian clustering procedures implemented in Structure v.2.3.4 [34].

Structure was run with three repetitions of 5×10^5 iterations following a burn-in period of 5×10^4 iterations, using the Admixture and Independent Allele Frequencies models [35], and assuming $K = 2$. As reference parental populations, we selected—from the ISPRA *Canis* database—the 39-STR genotypes of 190 unrelated wild individuals belonging to the Italian wolf population and 89 wolf-sized dogs living in rural areas [21]. Based on previous analyses performed on simulated wild, domestic, and hybrid genotypes, we assigned the unknown canid genotype to the Italian wolf population if its wolf membership proportion was $qi \geq 0.955$, to the dog population if $qi < 0.2$, otherwise it was considered as admixed for intermediate ($0.2 < qi < 0.954$) values (see Caniglia et al. [21] for further information about reference population choice and threshold selection).

A similar approach was used for the felid genotype, using as reference parental populations the 29-STR genotypes of 48 unrelated European wildcats, representative of the species distribution range in the central Apennines, and 65 free-ranging domestic cats, selected from the ISPRA *Felis* database [26]. However, when using another specific molecular marker set, such as the mentioned 29 domestic cat-derived STR panel [26,29,30], simulated genotype analyses suggested the application of different dedicated detection thresholds, assigning the unknown felid genotype to the European wildcat population if its wildcat membership proportion was $qi \geq 0.8$, to the domestic cat if $qi < 0.2$, whereas it was considered as admixed for intermediate ($0.2 < qi < 0.799$) values [11].

3. Results

The analysis of cuticle and medulla patterns morphologically classified the hair sample as belonging to *Felis* sp., suggesting it was a domestic cat, though genetic analyses on the same specimen returned a reliable *multilocus* genotype ($R > 95\%$) only at the wolf microsatellite panel with an average rate of missing data of 0.154. Conversely, the claw DNA sample yielded a reliable genotype ($R > 95\%$) only at the felid microsatellite panel with an average rate of missing data of 0.315. Mean rates of AS, ADO, and FA across loci were 0.846, 0, and 0 (standard errors: 0.056, 0, 0) for the hair sample and 0.638, 0.049 and 0.009 (standard errors: 0.083, 0.03, 0.009) for the claw sample, respectively. In the PCoA including canid reference populations, the hair genotype plotted marginally to reference wolves (Figure 1a), whereas in the PCoA including felid reference populations the claw genotype completely overlapped with reference domestic cats (Figure 1b). Assignment procedures clearly supported the outcomes from the PCoA, classifying the canid genotype obtained from the hair sample as belonging to an individual with slight genetic signs of dog ancestry, showing a proportion of posterior probability to belong to the wolf cluster $qi = 0.954$ (95% confidential interval C.I. = 0.78–1, Figure 1c). The *Amelogenin* analysis suggested it was a male (heterozygous genotype) individual with a Y-chromosome haplotype typical of the Italian wolf population [17], whereas the absence of the K-*locus* deletion theoretically indicated an animal with a wild-type coat colour. The claw genotype was unambiguously assigned to the domestic cat group with a proportion of posterior probability $qi = 0.996$ (95% C.I. = 0.982–1, Figure 1d).

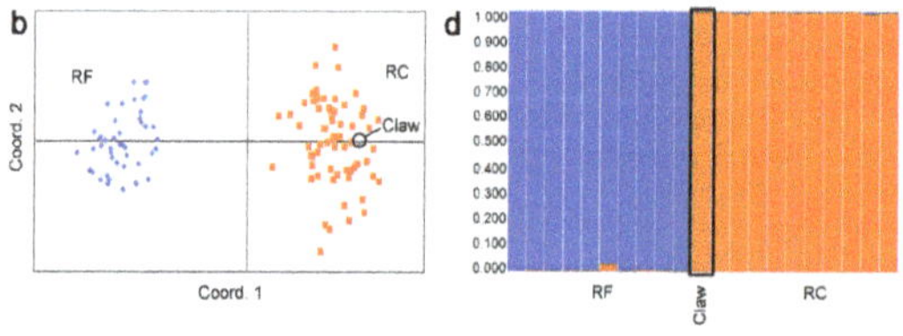

Figure 1. Multivariate and Bayesian assignment of *multilocus* genotypes obtained from the hair and claw samples to their respective belonging populations. Left side: graphical plotting of the Principal Coordinate Analysis performed in GenAlEx assigning (**a**) the hair sample genotype (grey triangle) to the cluster of the reference wolf (RW, blue) or dog (RD, orange) populations; (**b**) the claw sample genotype (grey triangle) to the cluster of the reference wildcat (RF, blue) or domestic cat (RC, orange) populations. Right side: Bayesian clustering histograms produced by Structure assuming $K = 2$ clusters and with the "admixture" and "I" models of the (**c**) hair genotype determined at 39 microsatellite *loci* and assigned to the reference wolf (RW, blue) or dog (RD, orange) populations and (**d**) claw genotype determined at 29 microsatellite *loci* and assigned to the reference wildcat (RF, blue) or domestic cat (RC, orange) populations. Each individual is represented by a vertical bar fragmented into the two-coloured sections, according to their proportion of membership to the reference population clusters. For a better readability, only ten genotypes for each reference population were shown as bar plots obtained from Bayesian assignment procedures.

4. Discussion

Non-invasive genetic sampling can provide useful insights into trophic relationships within ecosystems, clarifying fundamental ecological aspects such as feeding habits, predation behaviour, and dynamics of animal species, thus contributing to designing sound and long-term conservation strategies [36–38]. This approach is particularly when applied to elusive and threatened large or meso-carnivores such as the wolf and the European wildcat [23,39]. These two iconic species can somewhat share their ecological niches and sometimes their territories could overlap those of their domestic free-ranging relatives, increasing the risk of anthropogenic hybridisation and its possible deleterious consequences [40]. Such phenomenon, if widely spread, could undermine the gene pool integrity of wild ancestors through the introgression of domestic artificially selected genetic variants, which might potentially affect morphological, physiological, and behavioural traits of natural populations [13,41].

In this case study, we used the ecological, morphological, and genetic information contained in a biological canid faecal sample to detect any sign of trophic interactions between wolf and European wildcat individuals or their domestic counterparts. Though traditional molecular diet analyses are not able to identify with certainty the taxonomic status of the involved *taxa* below the specific level [7], we overcame this limitation by applying a multidisciplinary detection approach.

Firstly, we conducted traditional morphological analyses, which revealed the presence of felid remains in the collected canid faecal sample. Then, we used reliable molecular tools and highly discriminant canid and felid STR panels [12,26], which allowed us to reconstruct both the predator and prey genetic profiles. Finally, we exploited the availability of well-represented wild and domestic canid and felid reference populations to identify their correct taxonomic status at the subspecies level through Bayesian clustering analyses.

Interestingly, despite the initial water treatment of the faecal sample to separate its content, the few remaining canid intestinal cells on the hair sample allowed us to fully reconstruct the predator genetic profile with very high amplification success rates and no sign of ADO or FA across *loci*. Such a profile resulted in a wolf with traces of dog ancestry at autosomal microsatellite *loci*, consistently with the presence of wolf-dog admixture cases, repeatedly documented in this region using both video-trapping and genetic tools [13,42]. We were also able to reconstruct the genetic profile of the individual to which the claw belonged but with lower amplification success and higher error rates, which were, however, consistent with those reported in other non-invasive felid studies [16,20,43], probably due to the partial intestinal digestion the claw was exposed to. Such a felid profile resulted in a domestic cat and assessment of the problematic presence of free-ranging cat individuals in rural areas. This might increase the risk of anthropogenic admixture with wild animals as well as of predation on small vertebrates, thus bearing negative consequences for the conservation of biodiversity.

Sporadic cat consumption by wolves has been reported [44] and our findings might suggest a trophic interaction between admixed and domestic individuals into the wild. However, the analysis of food remains from scats cannot rule out the *post-mortem* consumption of a cat carcass. Although our results were obtained from a single case study, future genetic identifications from faecal samples combined with accurate detections of the consumed prey, if extensively applied, could provide additional useful data about possible shifts in food habits and habitat use between wild predators and their domestic counterparts. This is particularly useful in areas with a documented presence of wild and domestic overlapping [10,45], though to date significant different foraging strategies between wolf-dog admixed individuals and wolves have not been documented in the few available studies [10], and no similar data are currently reported for European wildcats and domestic cats.

Our multidisciplinary approach could also be applied to support future studies on possible behavioural alterations between wolves and wolf-dog hybrids in predatory strategies, contributing to detect the presence of admixture on one of their main preys, the wild boar, which is also facing anthropogenic hybridisation with the domestic pig [46,47]. However, results from this case study, although promising, are preliminary and should be confirmed by further studies based on larger sample sizes to statistically support the reliability of the applied method.

Our methodological procedures could be further improved by using highly performing markers to genotype non-invasively collected samples such as metabarcoding or highly informative and easily inter-lab comparable SNP panels, in cases of canid or felid predation on wildlife or other domestic pets to establish the real taxonomic status of the predators through a finer scale admixture identification and to provide a more reliable assessment of their potential impact on other threatened species [20,47–49].

5. Conclusions

Our results, although limited to the analysis of a single biological sample, clearly show how a multidisciplinary approach allows the cost-effective identification of both predator and prey *multilocus* genetic profiles and the accurate detection of their taxonomic status from non-invasive DNA. This will contribute to improve our knowledge about feeding habits, predatory dynamics, and anthropogenic hybridisation risk in threatened species, thus collecting information which is key to plan adequate conservation management actions.

Author Contributions: Conception and planning: E.V., F.M., L.L., E.F. and R.C.; Collection of non-invasive samples: L.L.; Morphological analysis: L.L., G.P. and I.B.; Genetic analysis: E.V., F.M. and R.C.; Statistical analyses: E.V., F.M., E.F. and R.C.; Writing—original draft preparation: E.V., F.M., L.L., E.F. and R.C.; Writing—review and editing E.V., F.M., L.L., E.F., N.M. and R.C. All authors have read and agreed to the published version of the manuscript.

Funding: This research received no external funding.

Institutional Review Board Statement: Not applicable.

Informed Consent Statement: Not applicable.

Data Availability Statement: Data presented in this study are available on request.

Acknowledgments: We are particularly grateful to all the colleagues who contributed in different years and projects to collect canid and felid biological samples, and gave invaluable advice and support, in particular: Ferretti F., Ragni B., De Faveri A., Giuliani A., Guberti V., Cagnolati G., Vercillo F., Di Croce A., Oliveira R., Tedaldi G., Di Masso E., Gentile L., Lapini L., Mallia E., DeLogu M., Scandura M., Mattioli L., Berzi D., Ciucci P., Apollonio M., Galaverni M., Cappai N., Pedrazzoli C, and Battilani D. We are also indebted to Selvatici M.A., Dotti R., Cazzato M., and Morreale E. (ISPRA) for their support with the administrative practices needed for the publication of the manuscript.

Conflicts of Interest: The authors declare no conflict of interest.

References

1. Trites, A.W.; Joy, R. Dietary Analysis from Fecal Samples: How Many Scats Are Enough? *J. Mammal.* **2005**, *86*, 704–712. [CrossRef]
2. Kartzinel, T.R.; Chen, P.A.; Coverdale, T.C.; Erickson, D.L.; Kress, W.J.; Kuzmina, M.L.; Rubenstein, D.I.; Wang, W.; Pringle, R.M. DNA Metabarcoding Illuminates Dietary Niche Partitioning by African Large Herbivores. *Proc. Natl. Acad. Sci. USA* **2015**, *112*, 8019–8024. [CrossRef] [PubMed]
3. Power, M.E.; Tilman, D.; Estes, J.A.; Menge, B.A.; Bond, W.J.; Mills, L.S.; Daily, G.; Castilla, J.C.; Lubchenco, J.; Paine, R.T. Challenges in the Quest for Keystones: Identifying Keystone Species Is Difficult—but Essential to Understanding How Loss of Species Will Affect Ecosystems. *Bioscience* **1996**, *46*, 609–620. [CrossRef]
4. Chetri, M.; Odden, M.; Wegge, P. Snow Leopard and Himalayan Wolf: Food Habits and Prey Selection in the Central Himalayas, Nepal. *PLoS ONE* **2017**, *12*, e0170549. [CrossRef] [PubMed]
5. Momeni, S.; Malekian, M.; Hemami, M.R. Molecular versus Morphological Approaches to Diet Analysis of the Caracal (*Caracal caracal*). *Mammalia* **2019**, *83*, 586–592. [CrossRef]
6. Shehzad, W.; Riaz, T.; Nawaz, M.A.; Miquel, C.; Poillot, C.; Shah, S.A.; Pompanon, F.; Coissac, E.; Taberlet, P. Carnivore Diet Analysis Based on Next-Generation Sequencing: Application to the Leopard Cat (*Prionailurus bengalensis*) in Pakistan. *Mol. Ecol.* **2012**, *21*, 1951–1965. [CrossRef] [PubMed]
7. Shi, Y.; Hoareau, Y.; Reese, E.; Wasser, S.K. Prey Partitioning between Sympatric Canid Species Revealed by DNA Metabarcoding. *bioRxiv* **2019**, *22*, 293–305. [CrossRef]

8. Pompanon, F.; Deagle, B.E.; Symondson, W.O.C.; Brown, D.S.; Jarman, S.N.; Taberlet, P. Who Is Eating What: Diet Assessment Using next Generation Sequencing. *Mol. Ecol.* **2012**, *21*, 1931–1950. [CrossRef]
9. Lee, O.; Lee, S.; Nam, D.H.; Lee, H.Y. Molecular Analysis for Investigating Dietary Habits: Genetic Screening of Prey Items In Scat And Stomach Contents Of Leopard Cats *Prionailurus bengalensis euptilurus*. *Zool. Stud.* **2013**, *52*, 45. [CrossRef]
10. Bassi, E.; Canu, A.; Firmo, I.; Mattioli, L.; Scandura, M.; Apollonio, M. Trophic Overlap between Wolves and Free-Ranging Wolf x Dog Hybrids in the Apennine Mountains, Italy. *Glob. Ecol. Conserv.* **2017**, *9*, 39–49. [CrossRef]
11. Randi, E.; Pierpaoli, M.; Beaumont, M.; Ragni, B.; Sforzi, A. Genetic Identification of Wild and Domestic Cats (*Felis silvestris*) and Their Hybrids Using Bayesian Clustering Methods. *Mol. Biol. Evol.* **2001**, *18*, 1679–1693. [CrossRef] [PubMed]
12. Randi, E.; Hulva, P.; Fabbri, E.; Galaverni, M.; Galov, A.; Kusak, J.; Bigi, D.; Bolfíková, B.Č.; Smetanová, M.; Caniglia, R.; et al. Multilocus Detection of Wolf x Dog Hybridization in Italy, and Guidelines for Marker Selection. *PLoS ONE* **2014**, *9*, e86409. [CrossRef] [PubMed]
13. Galaverni, M.; Caniglia, R.; Pagani, L.; Fabbri, E.; Boattini, A.; Randi, E. Disentangling Timing of Admixture, Patterns of Introgression, and Phenotypic Indicators in a Hybridizing Wolf Population. *Mol. Biol. Evol.* **2017**, *34*, 2324–2339. [CrossRef]
14. Mattucci, F.; Oliveira, R.; Lyons, L.A.; Alves, P.C.; Randi, E. European Wildcat Populations Are Subdivided into Five Main Biogeographic Groups: Consequences of Pleistocene Climate Changes or Recent Anthropogenic Fragmentation? *Ecol. Evol.* **2016**, *6*, 3–22. [CrossRef] [PubMed]
15. Galaverni, M.; Caniglia, R.; Fabbri, E.; Milanesi, P.; Randi, E. One, No One, or One Hundred Thousand: How Many Wolves Are There Currently in Italy? *Mammal Res.* **2016**, *61*, 13–24. [CrossRef]
16. Velli, E.; Bologna, M.A.; Silvia, C.; Ragni, B.; Randi, E. Non-Invasive Monitoring of the European Wildcat (*Felis silvestris silvestris* Schreber, 1777): Comparative Analysis of Three Different Monitoring Techniques and Evaluation of Their Integration. *Eur. J. Wildl. Res.* **2015**, *61*, 657–668. [CrossRef]
17. Caniglia, R.; Fabbri, E.; Galaverni, M.; Milanesi, P.; Randi, E. Non-invasive Sampling and Genetic Variability, Pack Structure, and Dynamics in an Expanding Wolf Population. *J. Mammal.* **2014**, *95*, 41–59. [CrossRef]
18. Imbert, C.; Caniglia, R.; Fabbri, E.; Milanesi, P.; Randi, E.; Serafini, M.; Torretta, E.; Meriggi, A. Why Do Wolves Eat Livestock?: Factors Influencing Wolf Diet in Northern Italy. *Biol. Conserv.* **2016**, *195*, 156–168. [CrossRef]
19. Lozano, J.; Malo, A.F. Conservation of The European Wildcat (*Felis silvestris*) In Mediterranean Environments: A Reassessment Of Current Threats. In *Mediterranean Ecosystems: Dynamics, Management and Conservation*; Williams, G.S., Ed.; Nova Science Publishers: Hauppauge, NY, USA, 2012; pp. 1–31.
20. Mattucci, F.; Galaverni, M.; Lyons, L.A.; Alves, P.C.; Randi, E.; Velli, E.; Pagani, L.; Caniglia, R. Genomic Approaches to Identify Hybrids and Estimate Admixture Times in European Wildcat Populations. *Sci. Rep.* **2019**, *9*, 11612. [CrossRef]
21. Caniglia, R.; Galaverni, M.; Velli, E.; Mattucci, F.; Canu, A.; Apollonio, M.; Mucci, N.; Scandura, M.; Fabbri, E. A Standardized Approach to Empirically Define Reliable Assignment Thresholds and Appropriate Management Categories in Deeply Introgressed Populations. *Sci. Rep.* **2020**, *10*, 2862. [CrossRef]
22. Ferretti, F.; Lovari, S.; Mancino, V.; Burrini, L.; Rossa, M. Food Habits of Wolves and Selection of Wild Ungulates in a Prey-Rich Mediterranean Coastal Area. *Mamm. Biol.* **2019**, *99*, 119–127. [CrossRef]
23. Figueiredo, A.M.; Valente, A.M.; Barros, T.; Carvalho, J.; Silva, D.A.M.; Fonseca, C.; de Carvalho, L.M.; Torres, R.T. What Does the Wolf Eat? Assessing the Diet of the Endangered Iberian Wolf (*Canis Lupus signatus*) in Northeast Portugal. *PLoS ONE* **2020**, *15*, e0230433. [CrossRef] [PubMed]
24. Caniglia, R.; Fabbri, E.; Greco, C.; Galaverni, M.; Randi, E. Forensic DNA against Wildlife Poaching: Identification of a Serial Wolf Killing in Italy. *Forensic Sci. Int.* **2010**, *1*, 331–338. [CrossRef]
25. Fabbri, E.; Velli, E.; D'Amico, F.; Galaverni, M.; Mastrogiuseppe, L.; Mattucci, F.; Caniglia, R. From Predation to Management: Monitoring Wolf Distribution and Understanding Depredation Patterns from Attacks on Livestock. *Hystrix Ital. J. Mammal.* **2018**, *29*, 101–110. [CrossRef]
26. Mattucci, F.; Oliveira, R.; Bizzarri, L.; Vercillo, F.; Anile, S.; Ragni, B.; Lapini, L.; Sforzi, A.; Alves, P.C.; Lyons, L.A.; et al. Genetic Structure of Wildcat (*Felis silvestris*) Populations in Italy. *Ecol. Evol.* **2013**, *3*, 2443–2458. [CrossRef]
27. Teerink, B.J. *Hair of West-European Mammals: Atlas and Identification Key*; Cambridge University Press: Cambridge, UK, 1991.
28. Taberlet, P. Reliable Genotyping of Samples with Very Low DNA Quantities Using PCR. *Nucleic Acids Res.* **1996**, *24*, 3189–3194. [CrossRef] [PubMed]
29. Menotti-Raymond, M.; O'Brien, S.J. Evolutionary Conservation of Ten Microsatellite Loci in Four Species of Felidae. *J. Hered.* **1995**, *86*, 319–322. [CrossRef]
30. MenottiRaymond, M.A.; David, V.A.; Obrien, S.J. Pet Cat Hair Implicates Murder Suspect. *Nature* **1997**, *386*, 774. [CrossRef]
31. Valière, N.; Valière, N. Gimlet: A Computer Program for Analysing Genetic Individual Identification Data. *Mol. Ecol. Notes* **2002**, *2*, 377–379. [CrossRef]
32. Miller, D.H.; Jensen, A.L.; Hammill, J.H. Density Dependent Matrix Model for Gray Wolf Population Projection. *Ecol. Modell.* **2002**, *151*, 271–278. [CrossRef]
33. Peakall, R.; Smouse, P.E. GenAlEx 6.5: Genetic Analysis in Excel. Population Genetic Software for Teaching and Research—An Update. *Bioinformatics* **2012**, *28*, 2537–2539. [CrossRef] [PubMed]
34. Pritchard, J.K.; Stephens, M.; Donnelly, P. Inference of Population Structure Using Multilocus Genotype Data. *Genetics* **2000**, *155*, 945–959. [CrossRef] [PubMed]

35. Falush, D.; Stephens, M.; Pritchard, J.K. Inference of Population Structure Using Multilocus Genotype Data: Linked Loci and Correlated Allele Frequencies. *Genetics* **2003**, *164*, 1567–1587. [CrossRef] [PubMed]
36. Schwartz, M.K.; Pilgrim, K.L.; McKelvey, K.S.; Rivera, P.T.; Ruggiero, L.F. DNA Markers for Identifying Individual Snowshoe Hares Using Field-Collected Pellets. *Northwest Sci.* **2007**, *81*, 316–322. [CrossRef]
37. Lukacs, P.M.; Eggert, L.S.; Burnham, K.P. Estimating Population Size from Multiple Detections with Non-Invasive Genetic Data. *Wildl. Biol. Pract.* **2007**, *3*, 83–92.
38. Thuo, D.; Furlan, E.; Broekhuis, F.; Kamau, J.; MacDonald, K.; Gleeson, D.M. Food from Faeces: Evaluating the Efficacy of Scat DNA Metabarcoding in Dietary Analyses. *PLoS ONE* **2019**, *15*, e0228950. [CrossRef]
39. Apostolico, F.; Vercillo, F.; La Porta, G.; Ragni, B. Long-Term Changes in Diet and Trophic Niche of the European Wildcat (*Felis silvestris silvestris*) in Italy. *Mammal Res.* **2016**, *61*, 109–119. [CrossRef]
40. Stronen, A.V.; Aspi, J.; Caniglia, R.; Fabbri, E.; Galaverni, M.; Godinho, R.; Kvist, L.; Mattucci, F.; Nowak, C.; von Thaden, A.; et al. Wolf-Dog Admixture Highlights the Need for Methodological Standards and Multidisciplinary Cooperation for Effective Governance of Wild x Domestic Hybrids. *Biol. Conserv.* **2022**, *266*, 109467. [CrossRef]
41. Todesco, M.; Pascual, M.A.; Owens, G.L.; Ostevik, K.L.; Moyers, B.T.; Heredia, S.M.; Hahn, M.A.; Caseys, C.; Bock, D.G. Hybridization and Extinction. *Evol. Appl.* **2016**, *9*, 892–908. [CrossRef]
42. Canu, A.; Mattioli, L.; Santini, A.; Apollonio, M.; Scandura, M. "Video-Scats": Combining Camera Trapping and Non-Invasive Genotyping to Assess Individual Identity and Hybrid Status in Gray Wolf. *Wildlife Biol.* **2017**, *2017*, 1–9. [CrossRef]
43. Anile, S.; Ragni, B.; Randi, E.; Mattucci, F.; Rovero, F. Wildcat Population Density on the Etna Volcano, Italy: A Comparison of Density Estimation Methods. *J. Zool.* **2014**, *293*, 252–261. [CrossRef]
44. Newsome, T.M.; Boitani, L.; Chapron, G.; Ciucci, P.; Dickman, C.R.; Dellinger, J.A.; Lopez-Bao, J.V.; Peterson, R.O.; Shores, C.R.; Wirsing, A.J.; et al. Food Habits of the World's Grey Wolves. *Mamm. Rev.* **2016**, *46*, 255–269. [CrossRef]
45. Germain, E.; Ruette, S.; Poulle, M.L. Likeness between the Food Habits of European Wildcats, Domestic Cats and Their Hybrids in France. *Mamm. Biol.* **2009**, *74*, 412–417. [CrossRef]
46. Scandura, M.; Iacolina, L.; Apollonio, M. Genetic Diversity In The European Wild Boar *Sus scrofa*: Phylogeography, Population Structure And Wild X Domestic Hybridization. *Mamm. Rev.* **2011**, *41*, 125–137. [CrossRef]
47. Lorenzini, R.; Fanelli, R.; Tancredi, F.; Siclari, A.; Garofalo, L. Matching STR and SNP Genotyping to Discriminate between Wild Boar, Domestic Pigs and Their Recent Hybrids for Forensic Purposes. *Sci. Rep.* **2020**, *10*, 3188. [CrossRef]
48. Dziech, A. Identification of Wolf-Dog Hybrids in Europe—An Overview of Genetic Studies. *Front. Ecol. Evol.* **2021**, *9*, 760160. [CrossRef]
49. Harmoinen, J.; von Thaden, A.; Aspi, J.; Kvist, L.; Cocchiararo, B.; Jarausch, A.; Gazzola, A.; Sin, T.; Lohi, H.; Hytönen, M.K.; et al. Reliable Wolf-Dog Hybrid Detection in Europe Using a Reduced SNP Panel Developed for Non-Invasively Collected Samples. *BMC Genom.* **2021**, *22*, 473. [CrossRef]

animals

MDPI

Review

Gut Microbiome Studies in Livestock: Achievements, Challenges, and Perspectives

Giovanni Forcina [1,2,3,*], Lucía Pérez-Pardal [1,2], Júlio Carvalheira [1,2,4] and Albano Beja-Pereira [1,2,5,6,*]

1. CIBIO, Centro de Investigação em Biodiversidade e Recursos Genéticos, *InBIO* Laboratório Associado, Campus de Vairão, Universidade do Porto, 4485-661 Vairão, Portugal
2. BIOPOLIS Program in Genomics, Biodiversity and Land Planning, CIBIO, Campus de Vairão, Universidade do Porto, 4485-661 Vairão, Portugal
3. Universidad de Alcalá, Global Change Ecology and Evolution Research Group (GloCEE), Departamento de Ciencias de la Vida, 28805 Alcalá de Henares, Spain
4. Abel Salazar Institute of Biomedical Sciences, University of Porto, Rua de Jorge Viterbo Ferreira 228, 4050-313 Porto, Portugal
5. DGAOT, Faculty of Sciences, Universidade do Porto, Rua Campo Alegre 687, 4169-007 Porto, Portugal
6. Sustainable Agrifood Production Research Centre (GreenUPorto), Universidade do Porto, Rua da Agrária 747, 4485-646 Vairão, Portugal

* Correspondence: giovanni.forcina@cibio.up.pt (G.F.); albanobp@cibio.up.pt (A.B.-P.); Tel.: +34-918854919 (G.F.)

Simple Summary: The relentless capacity of sequencing every bit of DNA at low cost has been fueling major advances in several research areas. This also applies to the animal sciences, which witnessed unprecedented progresses in fields such as animal nutrition, health, and breeding. Particular attention has been paid to the gut microbiome, the community of microorganisms inhabiting the digestive tract of livestock species, and unforeseen developments have arisen. Nonetheless, such efforts have not been equal for the different livestock species, and the vast majority rely on widely-used standard techniques through which taxonomically useful genetic data are generated rather than more informative—yet computationally demanding—organismal genome-wide variation data. This review offers a glimpse of the gut microbiome research on five emblematic livestock species touching on the limitations regarding (i) the major methodological frameworks, (ii) species or breed, (iii) and spatial reach of these studies, thus providing valuable indications to fill current knowledge gaps and hopefully lay the basis for the planning of concerted research efforts. In this respect, we conclude that future studies should extend shotgun sequencing and transcriptomic approaches primarily to largely neglected ovicaprine and chicken breeds from rural areas of developing countries and microbial groups other than bacteria.

Abstract: The variety and makeup of the gut microbiome are frequently regarded as the primary determinants of health and production performances in domestic animals. High-throughput DNA/RNA sequencing techniques (NGS) have recently gained popularity and permitted previously unheard-of advancements in the study of gut microbiota, particularly for determining the taxonomic composition of such complex communities. Here, we summarize the existing body of knowledge on livestock gut microbiome, discuss the state-of-the-art in sequencing techniques, and offer predictions for next research. We found that the enormous volumes of available data are biased toward a small number of globally distributed and carefully chosen varieties, while local breeds (or populations) are frequently overlooked despite their demonstrated resistance to harsh environmental circumstances. Furthermore, the bulk of this research has mostly focused on bacteria, whereas other microbial components such as protists, fungi, and viruses have received far less attention. The majority of these data were gathered utilizing traditional metabarcoding techniques that taxonomically identify the gut microbiota by analyzing small portions of their genome (less than 1000 base pairs). However, to extend the coverage of microbial genomes for a more precise and thorough characterization of microbial communities, a variety of increasingly practical and economical shotgun techniques are currently available.

Keywords: amplicon sequencing; genetic resources; metabarcoding; microbiota; next-generation sequencing; climate resilience; shotgun sequencing

1. Introduction

The massive decrease in sequencing costs associated with the generalization of high-throughput or next-generation sequencing (NGS) techniques has enabled unprecedented advances in microbiome studies spanning throughout the life sciences fields [1]. The strong bond between public health and the economy has been propelling the interest in microbiome research, which is deemed to hold a huge applicative potential under the One Health strategy [2] and other similar initiatives. Most of such studies addressing health and animal production have been mostly focused on gut microbiota, which is justified by the crucial role of these microorganisms in nutrition, fitness, and performance traits [3–5]. It is generally expected that advancing knowledge of the ruminant microbiome [6] bears a huge potential in terms of boosting animal production and health while lessening environmental pollution [7,8]. This promise seems of utmost importance when considering forecasts predicting an almost two-fold increase in the current production and consumption of meat in 30 years from now, with changes in dietary habits in developing countries—on top of human population growth—which will boost the demand for dairy products [9].

Some terms will be used extensively in this review and, for the benefit of the readers, their definition is provided as follows. The community of microorganisms inhabiting a given environment is referred to as a microbiota, while the term microbiome is used to indicate microbiota's collective genomes [10]. On the other hand, the concept of metagenomics, first defined as "the direct genetic analysis of genomes contained in an environmental sample" [11], has been elaborated further as the "study of the structure and function of entire nucleotide sequences isolated and analyzed from all the organisms (typically microbes) in a bulk sample" [12].

However, the fast-increasing body of research produced in the wake of the above-mentioned compelling socioeconomic reasons has yet to cover much ground. Among the main shortfalls of this kind of study is the almost exclusive focus on cosmopolitan and highly selected breeds, which lacks representativeness both in terms of diversity and functionality, as the most promising knowledge may come from locally adapted native breeds. The role of microorganisms in the resilience and performance of livestock species is of paramount interest for their potential commercial value, especially in a time of rampant global change.

Another critical factor is the uneven attention being paid to bacteria, which include taxa that cause significant economic losses in addition to being a serious hazard to public health, e.g., [13]. On the other hand, the other microorganisms, such as fungi, archaea, protozoa [14], and viruses, have received far less attention. Yet another weakness is that the classical metabarcoding approach is still largely used, as opposed to increasingly feasible and affordable shotgun approaches that are now available (although only for a low number of samples) for a more precise and extensive characterization of microbial communities.

The goal of this review, which was prompted by the steadily increasing number of articles devoted to the livestock gut microbiome, is to assess the primary body of research in this subject, provide an overview of the state-of-the-art regarding sequencing approaches and knowledge produced, and then offer suggestions for future studies.

2. Next-Generation Sequencing Techniques

2.1. Amplicon Metabarcoding

This technique, known as the large-scale taxonomic identification of biological samples through the analysis of short DNA fragments of one or more genes (known as DNA barcodes), has benefited significantly from the development of high-throughput sequencing, which made it possible to process complex environmental samples [https://metazoogene.org/metabarcoding (accessed on 10 September 2022)].

According to the taxonomic group being targeted, different barcodes are preferred. For example, the V2-V3 and V3-V4 *16S rRNA* hypervariable regions have being traditionally used for bacteria [15,16], the V4 and V9 *18S rRNA* hypervariable regions for protists [17], and the internal transcribed spacer (*ITS*) *rRNA* regions (*ITS1* and *ITS2*) for fungi [18,19]. This PCR-based method relies on a dual indexing mechanism to simultaneously process huge numbers of individual libraries (i.e., samples) covering many taxa at a low cost.

The operational taxonomic units (OTUs) characterized by means of these amplicon libraries, however, are an underrepresentation of the true microbial diversity in the community of interest because DNA barcoding has lower sensitivity and limited resolution when compared to metagenomic data (i.e., spanning entire genomes). Instead, PCR and sequencing errors may result in its overrepresentation [20]. Nevertheless, since amplicon metabarcoding has been widely used by the scientific community worldwide for almost 20 years, a large number of homologous sequences are available for free download from GenBank as well as from other widely used public repositories, including Greengenes v13_8 [21], SILVA 138 [22], and RDP18 (Ribosomal Database Project) [23].

However, it is important to note that these databases are not always regularly updated (the most recent updates were made in 2013, 2019, and 2020, respectively), which can be problematic for users. Another reason to rely on this locus over others is the accessibility of user-friendly software such as the Quantitative Insights into Microbial Ecology pipeline (QIIME) [24], which implements *16S*-based tools for taxonomic assignment. This even inspired the development of software that predicts functional profiles of bacterial populations based on their *16S* sequences, such as Tax4Fun [25]. Overall, this locus has been used in the majority of livestock microbiome investigations carried out until now, necessitating the establishment of recommendations and best practices for the benefit of the animal science community as a whole [26].

2.2. Shotgun Sequencing

This technique entails randomly shearing one or multiple genomes (for instance, in the case of an environmental sample) into small DNA fragments that are then individually sequenced, mapped to reference genomes and then reassembled in the proper order (for a quick explanation, see [27]). Such a technique, which is not based on PCR, has the benefit of avoiding the formation of amplification artifacts and, by being not reliant on taxon-specific primers, may produce more thorough and reliable results in terms of the overall microbial diversity associated with a given sample thanks to its high sensitivity and resolution power.

The provision of knowledge on the biological functions encoded by the genome(s) being sequenced is another significant benefit [28]. However, because of its high sequencing costs, this technique is not yet scalable for large-scale surveys based on high numbers of samples, despite being simple and fast to execute [29]. The difficulties in recreating the microbial composition in the case of complex and large communities and the high computing expenses connected with data storage and processing are additional limitations of this technique [28,29]. The rapidly expanding community of scientists using shotgun sequencing is fortunate to have access to powerful bioinformatics tools that are being made available, with some like BLAST+ [30] allowing the buildup of customized reference databases based on the inclusion of freely available nucleotide and protein sequences from public repositories.

2.3. Metatranscriptomics

Referred to as the study of genes that are transcribed in microbial communities at a given moment and under certain environmental conditions as measured by the abundance of collective RNA transcripts [31], this culture-independent approach has delivered major insights into niche-specific transcript expression patterns and the ecological functions of microbial taxa within their community [32]. Overall, a common drawback of this suite of techniques, especially RNASeq, is the high costs, which are nonetheless expected to drop in the coming years in parallel with an increase in computational power and specific software such as HISAT [33] or ABioTrans [34].

3. The Significance of Microbiome Studies in Livestock Species

Only five animal species—cows, chickens, goats, pigs, and sheep—produce the great majority of the animal products that humans consume (meat, milk, eggs) [35–37]. Each of these species has its own evolutionary history that can be very deep, as is the case of the chicken—a bird—when compared with the other four, which are mammals. Even among the latter, there are notable differences not only in their digestive tracts, but also in terms of physiological aspects such as growth and lifespan as well as reproduction and behavior [38]. There are significant differences between ruminants, which possess a multi-chambered stomach (consisting of reticulum, rumen, omasum and abomasum) used to digest plant materials through fermentation, and monogastric animals, whose stomach is a simple structure made of a single compartment. The advantages of the ruminants regarding their capacity to obtain energy from poor-quality food and the limitations experienced in maintaining a balanced and healthy ruminal flora do not apply to the monogastrics, which are characterized by a faster development and a shorter lifespan.

Indeed, the digestive tract is the most structural factor in animal production as its functionality and health determine most of the individual's performance [39], and is therefore the region that has been attracting the vast majority of microbiome research, followed by the reproductive tract [40]. Yet, more recently, environmental and public health concerns have also been the focus of a growing number of these studies, namely regarding greenhouse gas emissions [41], the spread of food-borne pathogens [42], and the rise of antibiotic resistance [43].

4. Microbiome Studies in Livestock Species

4.1. Ruminants

4.1.1. Cattle

The majority of gut microbiome studies in cattle have focused on the characterization of microbial communities by *16S rRNA* gene amplicon sequencing as a consequence of different animal diet composition [44,45], gastrointestinal tract (GIT) location [46], feed efficiency [47], breed-specificity [48], metabolic disturbs [49], changes over time [50] and individual specificities [51,52], as well as across housing types and farms [53]. Interestingly, special attention has been devoted to identifying individual-based differences irrespective of age, sex, breed, or environment [54], with patterns of similarity and dissimilarities helping to define the core microbiome in the bovine rumen [51] as well as other livestock [55]. At the same time, it was evidenced that differences in taxonomic composition and the underlying community metabolic networks may still result in functional similarity [56], as well as that the metabolic potential of the rumen microbiome may be diet-driven [57]. A large-scale survey of dairy cows indicated that the core rumen microbiome composition underlies not only animal productivity but also the nature of their emissions [58]. It was only recently that shotgun metagenomics opened the door to a thorough exploration of the rumen microbiome composition in cattle, enabling the assembly of entire bacterial genomes (most of which belong to new taxa), and the identification of new enzymes [59]. This approach has also allowed for the elucidation of the interplay between the rumen microbiome along with its metabolome and the host metabolome, shedding new light on the finest mechanisms underlying production performances in dairy cows [60].

4.1.2. Cattle Microbiome Profiling

Compared to other ruminant livestock species, the cattle gut microbiome is probably the one that has been explored more intensively, which provides an exhaustive picture of the bacterial communities inhabiting different GIT locations. The most abundant phyla are represented by Bacteroidetes and Firmicutes, which may account for more than 90%

of the entire GIT bacterial community, with Actinobacteria, Proteobacteria, Spirochaetes, and Tenericutes representing other major yet comparatively less abundant taxa [49,53,61–65]. Bacteroidetes and Firmicutes are dominated, respectively, by classes Bacterioidia and Clostridia along with Bacilli. Concerning the major orders (Figure 1), the former class mostly consists of Bacteroidales, while the latter one of Clostridiales [59]. The most abundant families include Bacteroidaceae, Clostridiaceae, Lachnospiraceae, Peptostreptococcaceae, Rikenellaceae, and Ruminococcaceae [53,55], while dominant genera—not only in cattle but in adult ruminants as a whole [55]—are *Butyrivibrio*, *Prevotella* and *Ruminococcus* [51,60,63,64,66]. Genus *Clostridium* is also abundant in cattle rumen [65] along with *Acetitomaculum*, *Acinetobacter*, *Mogibacterium*, *Succiniclasticum*, and *Treponema* [46]. Based on recent studies, genera like *Fibrobacter* and *Ruminococcus* are among the core heritable bacteria transferred vertically across generations in the light of their primary role in cellulolysis [58]. A detailed list of the GIT-associated bacterial taxa and the pertinent bibliographic references in cattle is reported in Table S1.

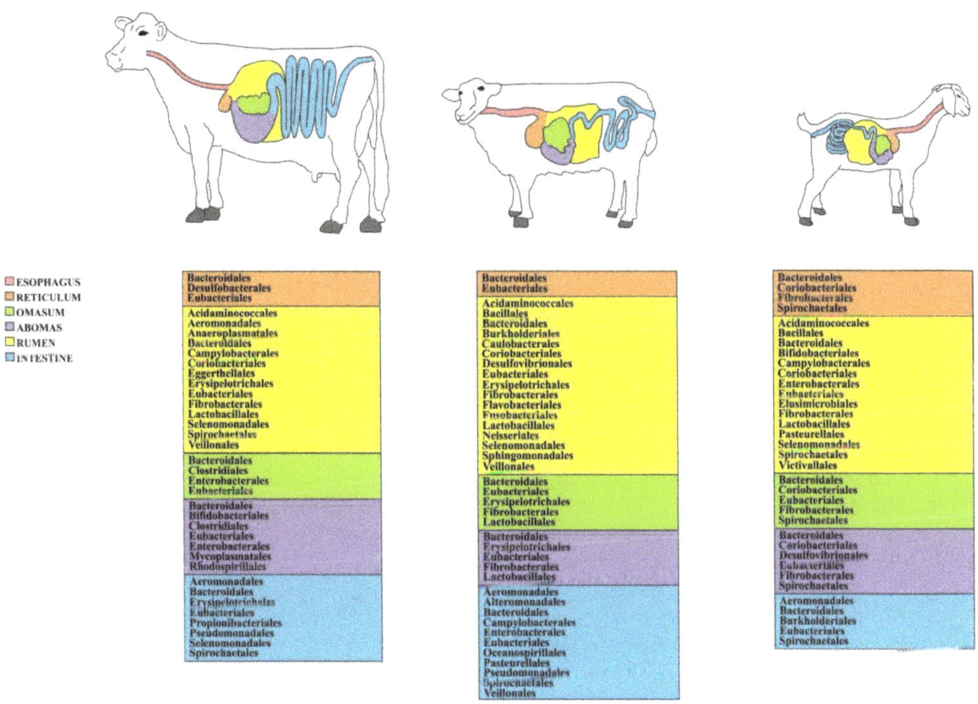

Figure 1. List of the most abundant bacterial orders found across the GITs of different ruminant livestock species (see Table S1 for further details). For the sake of clarity, the intestine designation may refer to both small and large intestines.

4.1.3. Sheep

The last decade has witnessed a mounting interest in sheep microbiome research. A recent study based on bacterial *16S* has confirmed that, similar to what was found in cows, the microbial hosts may be responsible for alterations in terms of feed efficiency [67], while other works have suggested that feeding strategies may promote a more or less diverse microbial community [68,69]. Additionally, compositional changes in the microbiome have been observed along the GIT [70,71] and as an effect of parasite infections [72]. In sheep, however, the compositional changes of the archaeal rather than the eubacterial community play a main role in feed efficiency, with the latter exerting its main influence in terms of the presence/absence pattern of only a few specific taxa [67]. Another recent *16S* study compared the microbiome composition in sheep and goats, finding no substantial differences between the two taxa; however, variation did occur depending on age, with older individuals hosting a higher microbial diversity [73], similar to what has been found in Tibetan sheep [71]. Interestingly, differences in gut bacterial compositions have been observed among different Chinese sheep breeds from the Tibetan Plateau [74], contradicting what was found in a similar study on Italian sheep where microbiome differences were mostly due to different husbandry practices [75]. Like in cattle, however, feed efficiency turned out to be related to a higher abundance and diversity of rumen microbiomes [76]. Other studies have been carried out on local breeds of high socioeconomic relevance, often revealing a fairly diverse composition, as in the case of the Chinese Mongolian sheep [77], or, similar to what was found in cattle and goats [78], a marked heterogeneity across different GIT locations as in the case of the Qinghai semi-fine wool sheep [71]. Notably, some recent studies addressing a likely association between host genetics and rumen microbiota in local sheep breeds have unveiled the modulating effect of ovine candidate genes on its composition [79] and the interplay between this and host gene expression in maintaining homeostasis in extreme environments [80]. Nevertheless, all the previous studies are based on *16S* metabarcoding, while applications of shotgun metagenomics to characterize the gut microbial composition in sheep are still scant. In this respect, however, it is worth mentioning a study combining the two approaches with metaproteomics to explore the link between microbial communities and biochemical pathways [81].

4.1.4. Sheep Microbiome Profiling

The characterization of the sheep GIT microbiome has revealed its substantial similarity in composition with that of cattle and other ruminants, with Bacteroidetes and Firmicutes making up more than 80 to 90 percent of the gut microbial community [67,69], followed by the phyla Actinobacteria, Proteobacteria, Spirochaetes, and Verrucomicrobia [68,74,82]. Bacterioidia and Clostridia are the dominant classes [75]. Moreover, Bacteroidales and Clostridiales figure among the most abundant orders (Figure 1), while, similar to what is observed in cattle, Eubacteriales and Lactobacillales stand out among Firmicutes. As far as the family-level is concerned, Ruminococcaceae and Lachnospiraceae emerge [74] along with Prevotellaceae, Rikenellaceae, and Succinivibrionaceae [67,76,80]. Concerning the most prevalent genera, *Prevotella* outstands [80], followed by *Acinetobacter* [79], *Campylobacter* [75], *Bacteroides, Desulfovibrio, Oscillospira, Ruminococcus, Treponema* [77], *Fibrobacter,* and *Succinivibrio* [76]. A detailed list of the GIT-associated bacterial taxa and the associated bibliographic references in sheep is provided in Table S1.

4.1.5. Goat

Molecular studies aimed at characterizing gut microbiome composition in this livestock species are still scarce in comparison to sheep, despite the economic relevance of goat meat and dairy products. Interesting exceptions, however, do occur, such as a work exploring the effects of dietary nitrate addition on microbial composition and ruminal fermentation based on a combined metabarcoding approach employing *16S* and *18S* amplicon libraries to characterize bacteria and protists along with fungi, respectively [83]. Other studies have evidenced the role played by fat acid supplementation [84] and a grain-rich diet [85] in shaping the bacterial and fungal diversity of rumen microbiome based on *16S* and *ITS* metabarcoding, respectively. Interestingly, a recent study based on amplicon libraries of the three loci mentioned before evidenced the role played by specific fungal and bacterial consortia in enabling lignocellulose breakdown by means of the production and interaction of a suite of specific metabolites [86]. Consistently, the *16S*-based methanogenic archaea diversity has turned out to be associated with a diet rich in condensed tannin-containing pine bark [87]. Current investigations have evidenced that, in goats as well, the microbial community varies throughout different GIT sectors [88] and tends to increase with age in young individuals [89,90], improving their productive performances [88,91,92]. Concordantly, the inoculation of rumen fluids during early life stages was found to boost the development of the rumen microbiome and even accelerate weaning [93], while the occurrence of apicomplexan parasites in goat kids was found to be associated with a decrease in the abundance of butyrate-producing bacteria, leading to an increase in mucosal inflammation and tissue repair [94]. Contrarily, it was discovered that antibiotic-induced gut microbiota dysbiosis likely worsened disease by encouraging inflammatory immune responses. [95].

Differences in the microbial composition have emerged when comparing adults belonging to different goat breeds [96], even if diet and environment seem to be the more important drivers of microbial diversity than genotype [97]. The occurrence of given bacterial hosts, in turn, was found to be associated with the digestibility of dietary phosphorus [98]. However, the exploration of gut microbiome components other than bacteria is quite limited in goats, with one of the few exceptions being represented by a study employing *16S* and *18S* amplicon libraries to explore the bacterial and ciliate protozoal diversity, respectively, in relation to the effects of antibacterial peptides on rumen fermentation function [99]. Moreover, the application of shotgun approaches to the characterization of the gut microbiome in goats is still limited to a single recent study [78].

4.1.6. Goat Microbiome Profiling

Compared to other ruminant livestock species, goats are probably those that have so far received less attention concerning gut microbiome studies. Bacteroidetes and Firmicutes are the dominant bacterial phyla (i.e., accounting for more than 80% of the GIT bacterial community), followed by Proteobacteria [89,90,98] and Verrumicrobia [84,88] along with Fibrobacteres, Spirochaetes, and Tenericutes [73,85]. As far as the most abundant orders are concerned, Bacteroidales and Clostridiales—similar to what is observed in cows and sheep—prevail over others (Figure 1) [96]. The dominant families include Prevotellaceae, Veillonellaceae, and, to a lesser extent, Lachnospiraceae, Rikenellaceae, and Ruminococcaceae [84,98]. Among the dominant genera, *Prevotella* stands out along with *Bacteroides, Butyrivibrio, Clostridium, Oscillospira, Ruminococcus, Succiniclasticum*, and *Succinivibrio* [73,84,85,88,90,92,96,100]. A list of the GIT-associated bacterial taxa and the related literature in goat is reported in Table S1.

4.2. Monogastric

4.2.1. Pig

Microbiome research in the pig industry has been propelled by the need to reduce animal stress that may otherwise turn into economic losses for farmers [101]. In this context, weaning is a critical life stage in which the piglet diet undergoes a sharp change. Studies on the swine gut microbiome have largely benefited from the establishment of a reference gene catalogue by means of deep metagenome sequencing of fecal samples [102] and have confirmed that also in this livestock species the interplay between diet and gut physiology across different growth stages is intimately associated with animal health and production performance [103], including fat deposition [104]. Other than varying on the basis of the food provided [105–108], GIT location [109–111], behavior [112], parasite infections [113], breed affiliation, and sex [114], the microbial diversity was found to correlate positively with piglet weight [115] and age [116]. Likewise, studies combining *16S rRNA* metabarcoding and shotgun metagenomic sequencing revealed that the composition of the pig gut microbiome varies considerably and predictably across the lifespan [117]. This is particularly evident postweaning [118], when a higher microbial diversity underlies an increase in the genes associated with oxidative stress and heat shock compared to nursing piglets [119]. Interestingly, some studies evidenced that the combination of culturomics and shotgun metagenomics—an approach seldom applied to other livestock species—may deliver a more exhaustive picture of gut [120,121] and antimicrobial resistance [122]. Investigations based on the combination of *16S rRNA* metabarcoding and shotgun metagenomics have delivered insights into antimicrobial resistance dynamics in pig farms [108], while *18S rRNA* metabarcoding of fecal samples allowed to draw up a detailed list of intestinal protist parasites [123]. The combination of *18S* and *ITS* amplicon libraries has been recently used to characterize the pig gut microbial eukaryote community, finding the association of some of its members with host body weight [124], while that of *16S* amplicon data and metagenomics has delivered unprecedented insights into the functional and taxonomic diversity of the pig gut microbiome [123].

4.2.2. Pig Microbiome Profiling

Notwithstanding the pronounced GIT structural differences between ruminants and monogastric animals such as pigs, the gut microbiome of the latter is also dominated by phyla Bacteroidetes and Firmicutes [115], followed by Proteobacteria [103,112], with Bacteroidia and Clostridia being the most abundant classes along with Bacilli [112,124]. Similar to what was observed in other livestock species, the dominant orders are Bacteroidales and Clostridiales (Figure 2), while the most abundant families are Bacteroidaceae, Enterobacteriaceae, Lachnospiraceae, Lactobacillaceae, Prevotellaceae, and Ruminococcaceae [106,108,116]. The genera most commonly found in the GIT of adult pigs are *Alloprevotella*, *Bacteroides*, *Escherichia*, *Lactobacillus*, and *Prevotella* [110,120,125,126] along with *Clostridium*, *Desulfovibrio*, *Enterococcus*, *Fusobacterium*, and *Streptococcus* [127–129]. A list of the GIT-associated bacterial taxa and the pertinent bibliographic references in pig is provided in Table S2.

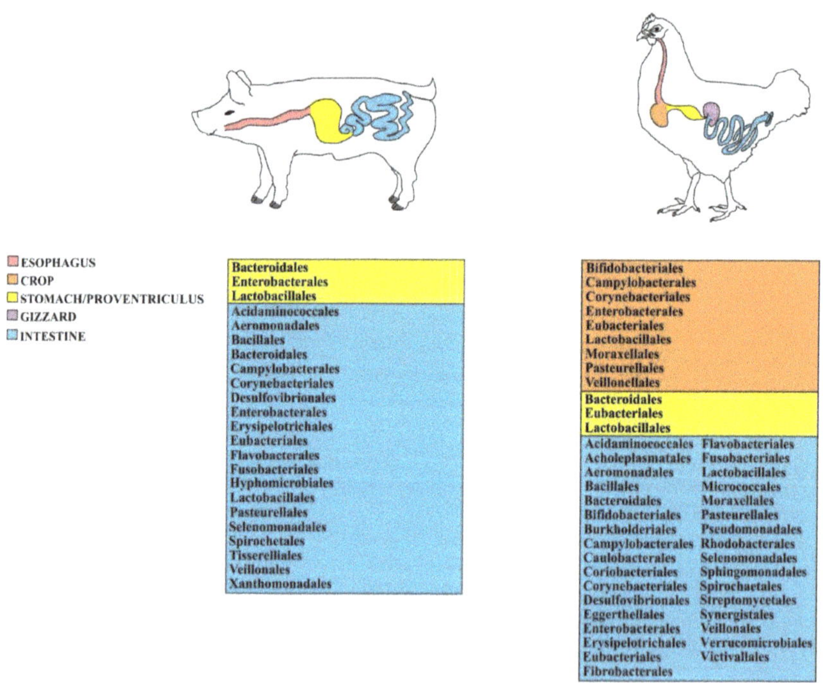

Figure 2. List of the most abundant bacterial orders found across the GITs of the two monogastric livestock species presented in this study (see Table S2 for further details).

4.2.3. Chicken

Microbiome research in chicken has made great strides since the advent of NGS techniques, as testified by the studies based on comparative metagenomic pyrosequencing to characterize the cecal microbiome in pathogen-free and infected individuals [130] and to explore the effect of antimicrobials on its communities as well as in relation to the abundance of antimicrobial resistance genes [131]. Nevertheless, most of these investigations are based on *16S rRNA* metabarcoding [132], while shotgun metagenomics is just taking its first steps in the poultry sector, with comparative studies applying the two approaches pointing to the much higher resolution power of the latter [133]. Shotgun metagenomics has also recently been employed to assess the role of dietary supplementation in improving the health status—and hence the productive performances—in broiler chickens by fostering the diversity of their cecum microbiome [134], also in the form of in ovo supplementation [92], as well as to characterize new bacterial, archaeal, and bacteriophage taxa of the chicken gut microbiome [135], thus shedding light on their biological function [136]. However, *16S rRNA* metabarcoding alone is still widely used to compare the microbiome composition of healthy versus unhealthy individuals as a consequence of viral, e.g., [137], or bacterial [138] infections, of individuals subjected to different dietary treatments [139], as well as across different indigenous breeds [140], GIT locations [141], rearing systems [142] and individual lifetimes [143], with a special focus on improving growth performance by transplanting cecal [144] or fecal [92] material between individuals of different age groups. However, compared to other livestock species, the non-bacterial component of the gut microbiome has been given more attention and most of the studies focus on possible pathogens such as *Cryptosporidium* [145].

4.2.4. Chicken Microbiome Profiling

Similar to what occurs in the GIT of other livestock species, the most abundant microbial phyla in chicken are Bacteroidetes and Firmicutes [136,146], even though sometimes Proteobacteria are more abundant than the former [135,138,145], while Bacilli, Clostridia, and Gammaproteobacteria are the dominant classes [134]. At the order level, Bacillales, Enterobacteriales, Lactobacillales, and Campylobacterales are the most common groups (Figure 2), while the most prevalent families include Enterobacteriaceae and Lactobacillaceae [139]. As far as the dominant genera are concerned, *Alistipes*, *Bacteroides*, *Clostridium*, *Helicobacter*, *Lactobacillus*, and *Ruminococcus* [133,143,144,146–149] stand out as well as *Flavobacterium* [139], *Campylobacter*, and *Veillonella* [150]. A detailed list of the GIT-associated bacterial taxa and the related bibliographic references in pig is reported in Table S2.

5. Resistome

The term "resistome" was introduced approximately two decades ago to indicate "the resistance determinants present in the soil" associated with bacterial populations living therein and showing multidrug resistance higher than expected [151]. The expression "bacterial resistome" has since become increasingly popular, while its meaning has evolved into the suite of all antibiotic resistance genes (ARGs) and their precursors in both pathogenic and nonpathogenic bacteria as well as antibiotic producers [152]. With a fast-growing body of research published on this topic, the concept of resistome has further evolved to incorporate different types of resistance and is now a key element in the framework of the One Health approach [153].

The identification of antimicrobial ARGs in bacteria inhabiting livestock GITs is crucial in animal science. An investigation specifically focused on the fecal bacterial resistome used a combination of the two approaches, traditional *16S* metabarcoding and shotgun metagenomics, evidencing the strong link between diet and antimicrobial resistance [154]. Moreover, the specificity of the microbial hosts in different GIT locations has emerged in a study on wild and domestic ungulates including cattle and goats [78]. This result serves as a model for future association research by highlighting the significance of local physiological changes along the GIT for various hosts. The advent of innovative nanopore technology, which enables large-scale research to highlight the most abundant resistance genes that may have a significant influence on animal, human, and environmental health, has spurred the rapidly expanding interest in the cow resistome [155]. On the other hand, studies on the bacterial resistome associated with the sheep gut microbiome are still scant when compared to those carried out in cattle or other livestock [156], and no specific investigation has been carried out on goats, but a recent study flagged as many as 30 ARGs in the sheep rumen, most of which related to daptomycin and colistin [157].

As far as non-ruminant livestock species are concerned, the scenario is even more complex. Pigs have received special attention in terms of characterization of the gut bacterial resistome, with recent studies demonstrating differential expression in humans, chickens, and specifically pigs [158]. Yet in chickens, the investigation into the ARGs associated with the gut microbiome has shown that the predominant classes are largely the same as those detected in pigs, including tetracycline, aminoglycoside and macrolide–lincosamide–streptogramin [159]. Of particular interest and utmost topicality is the risk of the potential transmission of ARGs from poultry meat to humans [160].

6. Metagenome and Functional Profile Prediction

Over the last years, a plethora of bioinformatics tools, including PICRUSt (Phylogenetic Investigation of Communities by Reconstruction of Unobserved States [161]), PICRUSt2 [162] along with FaproTax [163], the already mentioned Tax4Fun [25] and Tax4fun2 [164], have been made available to the scientific community for the purpose of predicting the functional profiles of the microbiota investigated in different studies. Moreover, Kyoto Encyclopedia of Genes and Genomes (KEGG) pathway analysis is often used in combination with these software to predict their metagenomic contributions. Since the vast majority of microbiome investigations have so far relied on *16S rRNA*, the algorithms of this type of program map the copies of this gene that were obtained in a given study to its homologs in the phylogenetically closest taxa with fully sequenced genomes. In other words, this approach allows predicting the functional metagenomic content without sequencing the entire genomes of the taxa which are actually present in the sample analyzed. Noteworthy, these software can work not only with amplicon metabarcoding but also shotgun sequencing data, even though their accuracy largely relies on available reference genomes and, as of now, it is still severely biased toward human datasets [165]. In addition, it is worth mentioning that a recent soil microbiome study comparing amplicon and shotgun sequencing functional profiling suggested that PICRUSt performs better than Tax4Fun to detect omnipresent functions, whereas Tax4Fun predicted greater abundances of functions from more specialized pathways [166]. Since the predictive tool used can lead to making different inferences, the authors suggested to reap the benefits of combining them rather than relying on either one or another [166].

The paucity of available reference genomes has been specifically invoked by some authors as the reason preventing them from performing functional prediction, e.g., [53], but others who nonetheless opted to perform it still detected significant differences in the predicted metagenomic profiles among GIT locations in dairy cows [46], sheep [77], goats [82], and pigs [126], thus pinpointing the most important metabolic pathways across different gastrointestinal microbial ecosystems. Additionally, functional prediction and KEGG analysis have been applied to unveil the differences in terms of metabolic pathways in the rumen of sheep cohorts with different feed efficiency [76] as well as in the same sheep sampled in different periods of the year [80], goats of different age [90], piglets with different body conditions [115], adult pigs of different breed and sex [71], and chickens of different age [167], breeds [140] or with a different health status [137]. It is conceivable that with the fast-growing increase in reference genomes available, metagenome and functional profile prediction tools will become more and more accurate in the next future, and their employment should be envisaged in any gut microbiome study.

7. Gut Microbiome, Health, and Welfare in Livestock

The positive or negative roles played by gut microbial taxa on health, welfare, behavior and performances in livestock exposed to different production conditions have been extensively discussed in several studies, e.g., [168–174]. It is worth touching on here why the studies addressing this topic are in such high demand nowadays and how crucial they are in animal farming as well as to broader society. In this respect, it is pertinent to mention their contribution to the development of non-antibiotic microbial therapies based on probiotics [175] as well as in pinpointing biomarkers of feed efficiency to deploy strategies that can notably improve livestock production performances [67,68,76,84,85,98,126,134,148,176] and growth [69,93,115]. Additionally, gut microbial profiling is paramount to monitor livestock health status and set up treatments to boost it, e.g., [113,125,139,144,165,177] as well as to prevent the establishment or aggravation of pathologic states [64,137,138,150] and evidence peculiar adaptations of local breeds to harsh environments [80].

8. Conclusions

The major advances in high-throughput sequencing technology have opened a new era in the study of the livestock gut microbiome, the composition and function of which are tightly associated with animal health and productive performance. The information produced has a profound social and economic impact. Previous attempts to take stock of available gut microbiome studies in livestock were mostly focused on cattle and chickens, or on the microbial groups rather than their hosts. In cattle, microbiome composition has been widely investigated in terms of feeding-related changes and their impacts on production strategies or environmental issues associated with ruminal methane emissions. We have expanded this review to other livestock animals, trying to make the point about what has been mostly performed so far and what is still lacking. A first consideration deals with the subjects of the microbiome studies carried out so far, in which priority was given to some species (such as cattle and pigs) rather than others (such as goats). Additionally, there is a clear bias in terms of the breeds investigated: expectedly, most studies are focused on a few cosmopolitan and highly selected breeds, while local breeds from rural areas are largely neglected, even though livestock research is a fundamental component to boost development strategies and the socioeconomic level of associated human communities. Characterizing the microbiome composition and its interaction with the host in non-intensive husbandry systems might, for instance, provide useful information on how to optimize livestock productivity through nutrient supplementation. Furthermore, it should be noted that a considerable number of microbiome studies, also on local breeds, have been carried out in China or Europe, while much less attention has been devoted to Africa and the tropical and subtropical regions as a whole.

Overall, the body of literature examined in this review allows us to conclude that livestock microbiome composition is affected by age, food, sex, and taxonomy, even if core bacteria occur across the GITs of different species. Admittedly, however, knowledge of other microbial groups is scant. As far as the methodological approach is concerned, shotgun metagenomics is still insipid when compared to amplicon metabarcoding sequencing, even though the few comparative studies employing both approaches on the same datasets evidenced the tremendously higher detection power of the former one, which is much more efficient in identifying underrepresented taxa whose detectability may be biased by the failure of universal primers to hybridize all templates as well as by its reliance on the number of hypervariable regions targeted, e.g., [133,178]. Conventional wisdom suggests that comparisons between studies based on either amplicon metabarcoding or shotgun sequencing on different datasets and with different experimental settings should not be made, but in general it can be stated that the latter is more reliable when estimating the absolute abundance of different microbial taxa. In addition, it is worth mentioning that in an increasing number of studies the two approaches are combined to first obtain a general picture of microbial diversity in the entire sample via amplicon metabarcoding and then, on this basis, select samples for shotgun sequencing to carry out functional analysis with higher accuracy, e.g., [108,127]. It is conceivable that with the fast-decreasing sequencing cost and the increasing suite of powerful bioinformatic tools available, the much more insightful shotgun sequencing will replace amplicon metabarcoding in most gut microbiome studies. This will presumably translate into expanding not only the taxonomic breadth and resolution but also the focus of the research. Indeed, bacteria is the most-studied microbial group compared to the others, which are most often given some attention only when represented by parasites of commercial relevance, but having a large amount of information encompassing other microbial groups as well may trigger interest in promoting research on them.

Indeed, focus on commensal protozoa, which nonetheless may still play a major role in regulating bacterial populations they feed on, is still limited, as is research on the fungal component of the gut microbiome. To achieve a comprehensive knowledge of the function of the microbiome and its underlying dynamics, the characterization of microbial groups other than bacteria is of key importance and should be addressed in future studies. As far as the NGS approach is concerned, it is important to note that the choice is often based on, other than budgetary issues, the availability and accessibility of comparative data as well as on the bioinformatics hurdles associated with the newest and most comprehensive techniques, which may prevent some research groups from applying them due to their limited computational resources and/or expertise. Enhancing the integration of meta-transcriptomic studies—which are particularly scant for non-bacterial components—into microbiome research would allow a better understanding of the functional role of different microbial groups in the gastrointestinal tract. Having such valuable tools should not deter researchers from embracing more exhaustive approaches such as those based on shotgun genomics or metatranscriptomics, which, on the one hand, are less affordable and more computationally intensive, yet, on the other hand, may deliver much larger and more accurate amounts of information. These efforts are fully justifiable if we consider that a major application of genomic data in relation to livestock studies is on animal and human health, where epidemiological investigations are fueled by the prospect of threats to human activity and public health with great impact on state wealth. Moreover, the integration of metagenomics and metatranscriptomics with metabolomics and proteomics (multi-omics sequencing) could provide more valuable information about the interaction of the complex "host-microbiota-environment", which could be useful for deploying future applications and interventions. In this context, there is a pressing need to better our understanding of the reciprocal influence of coexisting humans and livestock on each other's gut microbiome and resistome.

Supplementary Materials: The following supporting information can be downloaded at: https://www.mdpi.com/article/10.3390/ani12233375/s1, Table S1: List of most abundant eubacterial genera found across the GIT of ruminant livestock species; Table S2: List of most abundant eubacterial genera found across the GIT of monogastric livestock species.

Author Contributions: Conceptualization, A.B.-P.; methodology, G.F., L.P.-P.; validation, A.B.-P., J.C.; writing—original draft preparation, G.F.; writing—review and editing, G.F., L.P.-P. and A.B.-P.; visualization, G.F., L.P.-P.; supervision, A.B.-P., L.P.-P.; project administration, L.P.-P., J.C.; funding acquisition, L.P.-P., J.C. All authors have read and agreed to the published version of the manuscript.

Funding: This research was funded by the Portuguese Science and Technology foundation FCT/MCTES, through the grants PTDC/BAA-AGR/28866 and PTDC/BAA-AGR/028575, and from the European Regional Development Fund FEDER (contracts POCI-01-0145-FEDER-028866 and POCI-01-0145-FEDER-028575). LP-P holds a research contract from FCT/MCTES, while G.F. held a research contract through the project PTDC/BAA-AGR/28866 and is currently supported by the Ministry of Universities of the Spanish Government (María Zambrano/Next Generation EU). The article processing charge (APC) exemption was applied thanks to a waiver associated with the special issue "New Tools for Monitoring Genetic Diversity in Animals".

Institutional Review Board Statement: Not applicable.

Informed Consent Statement: Not applicable.

Data Availability Statement: Not applicable.

Conflicts of Interest: The authors declare no conflict of interest. The funders had no role in the design of the study; in the collection, analyses, or interpretation of data; in the writing of the manuscript; or in the decision to publish the results.

References

1. Amos, G.C.A.; Logan, A.; Anwar, S.; Fritzsche, M.; Mate, R.; Bleazard, T.; Rijpkema, S. Developing standards for the microbiome field. *Microbiome* **2020**, *8*, 98. [CrossRef] [PubMed]
2. European OneHealth/EcoHealth Workshop. Available online: http://www.biodiversity.be/health/58 (accessed on 20 May 2022).
3. Park, W. Gut microbiomes and their metabolites shape human and animal health. *J. Microbiol.* **2018**, *56*, 151–153. [CrossRef] [PubMed]
4. Ribeiro, G.O.; Gruninger, R.J.; Badhan, A.; McAllister, T.A. Mining the rumen for fibrolytic feed enzymes. *Anim. Front.* **2016**, *6*, 20–26. [CrossRef]
5. Weimer, P.J. Redundancy, resilience, and host specificity of the ruminal microbiota: Implications for engineering improved ruminal fermentations. *Front. Microbiol.* **2015**, *6*, 296. [CrossRef] [PubMed]
6. Malmuthuge, N.; Guan, L.L. Gut microbiome and omics: A new definition to ruminant production and health. *Anim. Front.* **2016**, *6*, 8–12. [CrossRef]
7. Alexander, T.W.; Plaizier, J.C. From the Editors: The importance of microbiota in ruminant production. *Anim. Front.* **2016**, *6*, 4–7. [CrossRef]
8. Cholewinska, P.; Czyz, K.; Nowakowski, P.; Wyrostek, A. The microbiome of the digestive system of ruminants—A review. *Anim. Health Res. Rev.* **2020**, *21*, 3–14. [CrossRef]
9. Alexandratos, N.; Bruinsma, J. *World Agriculture towards 2030/2050: The 2012 Revision*; ESA Working Paper 12-03; FAO: Rome, Italy, 2012.
10. Berg, G.; Rybakova, D.; Fischer, D.; Cernava, T.; Verges, M.C.; Charles, T.; Chen, X.; Cocolin, L.; Eversole, K.; Corral, G.H.; et al. Microbiome definition re-visited: Old concepts and new challenges. *Microbiome* **2020**, *8*, 103. [CrossRef]
11. Thomas, T.; Gilbert, J.; Meyer, F. Metagenomics—A guide from sampling to data analysis. *Microb. Inform. Exp.* **2012**, *2*, 3. [CrossRef]
12. National Human Genome Research Institute. Metagenomics. Available online: https://www.genome.gov/genetics-glossary/Metagenomics (accessed on 12 June 2022).
13. Wareth, G.; El-Diasty, M.; Abdel-Hamid, N.H.; Holzer, K.; Hamdy, M.E.R.; Moustafa, S.; Shahein, M.A.; Melzer, F.; Beyer, W.; Pletz, M.W.; et al. Molecular characterization and antimicrobial susceptibility testing of clinical and non-clinical *Brucella melitensis* and *Brucella abortus* isolates from Egypt. *One Health* **2021**, *13*, 100255. [CrossRef]
14. Hristov, A.N.; Ivan, M.; Rode, L.M.; McAllister, T.A. Fermentation characteristics and ruminal ciliate protozoal populations in cattle fed medium- or high-concentrate barley-based diets. *J. Anim. Sci.* **2001**, *79*, 515–524. [CrossRef] [PubMed]
15. Drancourt, M.; Bollet, C.; Carlioz, A.; Martelin, R.; Gayral, J.-P.; Raoult, R. 16S ribosomal DNA sequence analysis of a large collection of environmental and clinical unidentifiable bacterial isolates. *J. Clin. Microbiol.* **2000**, *38*, 3623–3630. [CrossRef]
16. Yang, B.; Wang, Y.; Qian, P.Y. Sensitivity and correlation of hypervariable regions in 16S rRNA genes in phylogenetic analysis. *BMC Bioinform.* **2016**, *17*, 135. [CrossRef]
17. Choi, J.; Park, J.S. Comparative analyses of the V4 and V9 regions of 18S rDNA for the extant eukaryotic community using the Illumina platform. *Sci. Rep.* **2020**, *10*, 6519. [CrossRef] [PubMed]
18. Bellemain, E.; Carlsen, T.; Brochmann, C.; Coissac, E.; Taberlet, P.; Kauserud, H. ITS as an environmental DNA barcode for fungi: An in silico approach reveals potential PCR biases. *BMC Microbiol.* **2010**, *10*, 189. [CrossRef] [PubMed]
19. Koljalg, U.; Nilsson, R.H.; Abarenkov, K.; Tedersoo, L.; Taylor, A.F.; Bahram, M.; Bates, S.T.; Bruns, T.D.; Bengtsson-Palme, J.; Callaghan, T.M.; et al. Towards a unified paradigm for sequence-based identification of fungi. *Mol. Ecol.* **2013**, *22*, 5271–5277. [CrossRef]
20. Poretsky, R.; Rodriguez, R.L.; Luo, C.; Tsementzi, D.; Konstantinidis, K.T. Strengths and limitations of 16S rRNA gene amplicon sequencing in revealing temporal microbial community dynamics. *PLoS ONE* **2014**, *9*, e93827. [CrossRef]
21. DeSantis, T.Z.; Hugenholtz, P.; Larsen, N.; Rojas, M.; Brodie, E.L.; Keller, K.; Huber, T.; Dalevi, D.; Hu, P.; Andersen, G.L. Greengenes, a chimera-checked 16S rRNA gene database and workbench compatible with ARB. *Appl. Environ. Microbiol.* **2006**, *72*, 5069–5072. [CrossRef]
22. Quast, C.; Pruesse, E.; Yilmaz, P.; Gerken, J.; Schweer, T.; Yarza, P.; Peplies, J.; Glockner, F.O. The SILVA ribosomal RNA gene database project: Improved data processing and web-based tools. *Nucleic Acids Res.* **2013**, *41*, D590–D596. [CrossRef]
23. Maidak, B.L.; Olsen, G.J.; Larsen, N.; Overbeek, R.; McCaughey, M.J.; Woese, C.R. The RDP (Ribosomal Database Project). *Nucleic Acids Res.* **1997**, *25*, 109–111. [CrossRef]
24. Bolyen, E.; Rideout, J.R.; Dillon, M.R.; Bokulich, N.A.; Abnet, C.C.; Al-Ghalith, G.A.; Alexander, H.; Alm, E.J.; Arumugam, M.; Asnicar, F.; et al. Reproducible, interactive, scalable and extensible microbiome data science using QIIME 2. *Nat. Biotechnol.* **2019**, *37*, 852–857. [CrossRef] [PubMed]
25. Asshauer, K.P.; Wemheuer, B.; Daniel, R.; Meinicke, P. Tax4Fun: Predicting functional profiles from metagenomic 16S rRNA data. *Bioinformatics* **2015**, *31*, 2882–2884. [CrossRef] [PubMed]
26. Weinroth, M.D.; Belk, A.D.; Dean, C.; Noyes, N.; Dittoe, D.K.; Rothrock, M.J.; Ricke, S.C.; Myer, P.R.; Henniger, M.T.; Ramirez, G.A.; et al. Considerations and best practices in animal science 16S ribosomal RNA gene sequencing microbiome studies. *J. Anim. Sci.* **2022**, *100*, skab346. [CrossRef] [PubMed]
27. National Human Genome Research Institute. Shotgun Sequencing. Available online: https://www.genome.gov/genetics-glossary/Shotgun-Sequencing (accessed on 12 June 2022).

28. Sharpton, T.J. An introduction to the analysis of shotgun metagenomic data. *Front. Plant. Sci.* **2014**, *5*, 209. [CrossRef]
29. Peabody, M.A.; Van Rossum, T.; Lo, R.; Brinkman, F.S. Evaluation of shotgun metagenomics sequence classification methods using in silico and in vitro simulated communities. *BMC Bioinform.* **2015**, *16*, 363. [CrossRef]
30. Camacho, C.; Coulouris, G.; Avagyan, V.; Ma, N.; Papadopoulos, J.; Bealer, K.; Madden, T.L. BLAST+: Architecture and applications. *BMC Bioinform.* **2009**, *10*, 421. [CrossRef]
31. O'Malley, M.A. Metatranscriptomics. In *Encyclopedia of Systems Biology*; Dubitzky, W., Wolkenhauer, O., Cho, K.H., Yokota, H., Eds.; Springer: New York, NY, USA, 2013.
32. Waikel, R.L.; Waikel, P.A. Metatranscriptomics. *AccessScience*. 2011. Available online: https://www.accessscience.com/content/article/aYB110044 (accessed on 27 November 2022).
33. Kim, D.; Langmead, B.; Salzberg, S.L. HISAT: A fast spliced aligner with low memory requirements. *Nat. Methods* **2015**, *12*, 357–360. [CrossRef]
34. Zou, Y.; Bui, T.T.; Selvarajoo, K. ABioTrans: A Biostatistical Tool for Transcriptomics Analysis. *Front. Genet.* **2019**, *10*, 499. [CrossRef]
35. FAO. How to feed the world in 2050. In *Proceedings of the Expert Meeting on How to Feed the World in 2050*; Food and Agriculture Organization of the United Nations: Rome, Italy, 2009.
36. FAO. FAOSTAT. Available online: http://www.fao.org/faostat/en/#home (accessed on 18 March 2022).
37. Pulina, G.; Milan, M.J.; Lavin, M.P.; Theodoridis, A.; Morin, E.; Capote, J.; Thomas, D.L.; Francesconi, A.H.D.; Caja, G. Invited review: Current production trends, farm structures, and economics of the dairy sheep and goat sectors. *J. Dairy Sci.* **2018**, *101*, 6715–6729. [CrossRef]
38. Reece, W.O. *Functional Anatomy and Physiology of Domestic Animals*, 4th ed.; Wiley-Blackwell: Ames, IA, USA, 2009.
39. Celi, P.; Cowieson, A.J.; Fru-Nji, F.; Steinert, R.E.; Kluenter, A.M.; Verlhac, V. Gastrointestinal functionality in animal nutrition and health: New opportunities for sustainable animal production. *Anim. Feed Sci. Technol.* **2017**, *234*, 88–100. [CrossRef]
40. Heil, B.A.; Paccamonti, D.L.; Sones, J.L. Role for the mammalian female reproductive tract microbiome in pregnancy outcomes. *Physiol. Genom.* **2019**, *51*, 390–399. [CrossRef] [PubMed]
41. Min, B.R.; Solaiman, S.; Waldrip, H.M.; Parker, D.; Todd, R.W.; Brauer, D. Dietary mitigation of enteric methane emissions from ruminants: A review of plant tannin mitigation options. *Anim. Nutr.* **2020**, *6*, 231–246. [CrossRef] [PubMed]
42. Black, Z.; Balta, I.; Black, L.; Naughton, P.J.; Dooley, J.S.G.; Corcionivoschi, N. The Fate of Foodborne Pathogens in Manure Treated Soil. *Front. Microbiol.* **2021**, *12*, 781357. [CrossRef]
43. Zalewska, M.; Blazejewska, A.; Czapko, A.; Popowska, M. Antibiotics and Antibiotic Resistance Genes in Animal Manure—Consequences of Its Application in Agriculture. *Front. Microbiol.* **2021**, *12*, 610656. [CrossRef] [PubMed]
44. Kim, M.; Kim, J.; Kuehn, L.A.; Bono, J.L.; Berry, E.D.; Kalchayanand, N.; Freetly, H.C.; Benson, A.K.; Wells, J.E. Investigation of bacterial diversity in the feces of cattle fed different diets. *J. Anim. Sci.* **2014**, *92*, 683–694. [CrossRef] [PubMed]
45. Palumbo, F.; Squartini, A.; Barcaccia, G.; Macolino, S.; Pornaro, C.; Pindo, M.; Sturaro, E.; Ramanzin, M. A multi-kingdom metabarcoding study on cattle grazing Alpine pastures discloses intra-seasonal shifts in plant selection and faecal microbiota. *Sci. Rep.* **2021**, *11*, 889. [CrossRef]
46. Mao, S.; Zhang, M.; Liu, J.; Zhu, W. Characterising the bacterial microbiota across the gastrointestinal tracts of dairy cattle: Membership and potential function. *Sci. Rep.* **2015**, *5*, 16116. [CrossRef]
47. McGovern, E.; Kenny, D.A.; McCabe, M.S.; Fitzsimons, C.; McGee, M.; Kelly, A.K.; Waters, S.M. 16S rRNA Sequencing Reveals Relationship Between Potent Cellulolytic Genera and Feed Efficiency in the Rumen of Bulls. *Front. Microbiol.* **2018**, *9*, 1842. [CrossRef]
48. Gonzalez-Recio, O.; Zubiria, I.; Garcia-Rodriguez, A.; Hurtado, A.; Atxaerandio, R. Short communication: Signs of host genetic regulation in the microbiome composition in 2 dairy breeds: Holstein and Brown Swiss. *J. Dairy Sci.* **2018**, *101*, 2285–2292. [CrossRef]
49. Plaizier, J.C.; Li, S.; Tun, H.M.; Khafipour, E. Nutritional Models of Experimentally-Induced Subacute Ruminal Acidosis (SARA) Differ in Their Impact on Rumen and Hindgut Bacterial Communities in Dairy Cows. *Front. Microbiol.* **2016**, *7*, 2128. [CrossRef]
50. Rudi, K.; Moen, B.; Sekelja, M.; Frisli, T.; Lee, M.R. An eight-year investigation of bovine livestock fecal microbiota. *Vet. Microbiol.* **2012**, *160*, 369–377. [CrossRef] [PubMed]
51. Jami, E.; Mizrahi, I. Composition and similarity of bovine rumen microbiota across individual animals. *PLoS ONE* **2012**, *7*, e33306. [CrossRef] [PubMed]
52. Li, R.W.; Connor, E.E.; Li, C.; Baldwin Vi, R.L.; Sparks, M.E. Characterization of the rumen microbiota of pre-ruminant calves using metagenomic tools. *Environ. Microbiol.* **2012**, *14*, 129–139. [CrossRef] [PubMed]
53. Hagey, J.V.; Bhatnagar, S.; Heguy, J.M.; Karle, B.M.; Price, P.L.; Meyer, D.; Maga, E.A. Fecal Microbial Communities in a Large Representative Cohort of California Dairy Cows. *Front. Microbiol.* **2019**, *10*, 1093. [CrossRef] [PubMed]
54. Durso, L.M.; Harhay, G.P.; Smith, T.P.; Bono, J.L.; Desantis, T.Z.; Harhay, D.M.; Andersen, G.L.; Keen, J.E.; Laegreid, W.W.; Clawson, M.L. Animal-to-animal variation in fecal microbial diversity among beef cattle. *Appl. Environ. Microbiol.* **2010**, *76*, 4858–4862. [CrossRef] [PubMed]
55. Henderson, G.; Cox, F.; Ganesh, S.; Jonker, A.; Young, W.; Global Rumen Census Collaborators; Janssen, P.H. Rumen microbial community composition varies with diet and host, but a core microbiome is found across a wide geographical range. *Sci. Rep.* **2015**, *5*, 14567. [CrossRef]

56. Taxis, T.M.; Wolff, S.; Gregg, S.J.; Minton, N.O.; Zhang, C.; Dai, J.; Schnabel, R.D.; Taylor, J.F.; Kerley, M.S.; Pires, J.C.; et al. The players may change but the game remains: Network analyses of ruminal microbiomes suggest taxonomic differences mask functional similarity. *Nucleic Acids Res.* **2015**, *43*, 9600–9612. [CrossRef]
57. Brulc, J.M.; Antonopoulos, D.A.; Miller, M.E.; Wilson, M.K.; Yannarell, A.C.; Dinsdale, E.A.; Edwards, R.E.; Frank, E.D.; Emerson, J.B.; Wacklin, P.; et al. Gene-centric metagenomics of the fiber-adherent bovine rumen microbiome reveals forage specific glycoside hydrolases. *Proc. Natl. Acad. Sci. USA* **2009**, *106*, 1948–1953. [CrossRef]
58. Wallace, R.J.; Sasson, G.; Garnsworthy, P.C.; Tapio, I.; Gregson, E.; Bani, P.; Huhtanen, P.; Bayat, A.R.; Strozzi, F.; Biscarini, F.; et al. A heritable subset of the core rumen microbiome dictates dairy cow productivity and emissions. *Sci. Adv.* **2019**, *5*, eaav8391. [CrossRef]
59. Stewart, R.D.; Auffret, M.D.; Warr, A.; Walker, A.W.; Roehe, R.; Watson, M. Compendium of 4941 rumen metagenome-assembled genomes for rumen microbiome biology and enzyme discovery. *Nat. Biotechnol.* **2019**, *37*, 953–961. [CrossRef]
60. Xue, M.Y.; Sun, H.Z.; Wu, X.H.; Liu, J.X.; Guan, L.L. Multi-omics reveals that the rumen microbiome and its metabolome together with the host metabolome contribute to individualized dairy cow performance. *Microbiome* **2020**, *8*, 64. [CrossRef]
61. Khafipour, E.; Li, S.; Plaizier, J.C.; Krause, D.O. Rumen microbiome composition determined using two nutritional models of subacute ruminal acidosis. *Appl. Environ. Microbiol.* **2009**, *75*, 7115–7124. [CrossRef] [PubMed]
62. Petri, R.M.; Schwaiger, T.; Penner, G.B.; Beauchemin, K.A.; Forster, R.J.; McKinnon, J.J.; McAllister, T.A. Characterization of the Core Rumen Microbiome in Cattle during Transition from Forage to Concentrate as Well as during and after an Acidotic Challenge. *PLoS ONE* **2013**, *8*, e83424. [CrossRef] [PubMed]
63. Jami, E.; Israel, A.; Kotser, A.; Mizrahi, I. Exploring the bovine rumen bacterial community from birth to adulthood. *ISME J.* **2013**, *7*, 1069–1079. [CrossRef] [PubMed]
64. Mao, S.; Zhang, R.; Wang, D.; Zhu, W. Impact of subacute ruminal acidosis (SARA) adaptation on rumen microbiota in dairy cattle using pyrosequencing. *Anaerobe* **2013**, *24*, 12–19. [CrossRef] [PubMed]
65. Kim, Y.H.; Nagata, R.; Ohkubo, A.; Ohtani, N.; Kushibiki, S.; Ichijo, T.; Sato, S. Changes in ruminal and reticular pH and bacterial communities in Holstein cattle fed a high-grain diet. *BMC Vet. Res.* **2018**, *14*, 310. [CrossRef]
66. Myer, P.; Freetly, H.; Wells, J.; Smith, T.; Kuehn, L. Analysis of the gut bacterial communities in beef cattle and their association with feed intake, growth, and efficiency. *J. Anim. Sci.* **2017**, *95*, 3215–3224. [CrossRef]
67. McLoughlin, S.; Spillane, C.; Claffey, N.; Smith, P.E.; O'Rourke, T.; Diskin, M.G.; Waters, S.M. Rumen Microbiome Composition Is Altered in Sheep Divergent in Feed Efficiency. *Front. Microbiol.* **2020**, *11*, 1981. [CrossRef]
68. Fu, Z.; Xu, X.; Zhang, J.; Zhang, L. Effect of different feeding methods on rumen microbes in growing Chinese Tan sheep. *Rev. Bras. Zootec.* **2020**, *49*, e20190258. [CrossRef]
69. Yu, S.; Zhang, G.; Liu, Z.; Wu, P.; Yu, Z.; Wang, J. Repeated inoculation with fresh rumen fluid before or during weaning modulates the microbiota composition and co-occurrence of the rumen and colon of lambs. *BMC Microbiol.* **2020**, *20*, 29. [CrossRef]
70. Wang, J.; Fan, H.; Han, Y.; Zhao, J.; Zhou, Z. Characterization of the microbial communities along the gastrointestinal tract of sheep by 454 pyrosequencing analysis. *Asian-Australas. J. Anim. Sci.* **2017**, *30*, 100–110. [CrossRef] [PubMed]
71. Wang, X.; Hu, L.; Liu, H.; Xu, T.; Zhao, N.; Zhang, X.; Geng, Y.; Kang, S.; Xu, S. Characterization of the bacterial microbiota across the different intestinal segments of the Qinghai semi-fine wool sheep on the Qinghai-Tibetan Plateau. *Anim. Biosci.* **2021**, *34*, 1921–1929. [CrossRef] [PubMed]
72. Cortes, A.; Wills, J.; Su, X.; Hewitt, R.E.; Robertson, J.; Scotti, R.; Price, D.R.G.; Bartley, Y.; McNeilly, T.N.; Krause, L.; et al. Infection with the sheep gastrointestinal nematode *Teladorsagia circumcincta* increases luminal pathobionts. *Microbiome* **2020**, *8*, 60. [CrossRef] [PubMed]
73. Shabana, I.I.; Albakri, N.N.; Bouqellah, N.A. Metagenomic investigation of faecal microbiota in sheep and goats of the same ages. *J. Taibah Univ. Sci.* **2020**, *15*, 1–9. [CrossRef]
74. Chang, J.; Yao, X.; Zuo, C.; Qi, Y.; Chen, D.; Ma, W. The gut bacterial diversity of sheep associated with different breeds in Qinghai province. *BMC Vet. Res.* **2020**, *16*, 254. [CrossRef]
75. Minozzi, G.; Biscarini, F.; Dalla Costa, E.; Chincarini, M.; Ferri, N.; Palestrini, C.; Minero, M.; Mazzola, S.; Piccinini, R.; Vignola, G.; et al. Analysis of Hindgut Microbiome of Sheep and Effect of Different Husbandry Conditions. *Animals* **2020**, *11*, 4. [CrossRef]
76. Zhang, Y.K.; Zhang, X.X.; Li, F.D.; Li, C.; Li, G.Z.; Zhang, D.Y.; Song, Q.Z.; Li, X.L.; Zhao, Y.; Wang, W.M. Characterization of the rumen microbiota and its relationship with residual feed intake in sheep. *Animal* **2021**, *15*, 100161. [CrossRef]
77. Zeng, Y.; Zeng, D.; Ni, X.; Zhu, H.; Jian, P.; Zhou, Y.; Xu, S.; Lin, Y.; Li, Y.; Yin, Z.; et al. Microbial community compositions in the gastrointestinal tract of Chinese Mongolian sheep using Illumina MiSeq sequencing revealed high microbial diversity. *AMB Express* **2017**, *7*, 75. [CrossRef]
78. Xie, F.; Jin, W.; Si, H.; Yuan, Y.; Tao, Y.; Liu, J.; Wang, X.; Yang, C.; Li, Q.; Yan, X.; et al. An integrated gene catalog and over 10,000 metagenome-assembled genomes from the gastrointestinal microbiome of ruminants. *Microbiome* **2021**, *9*, 137. [CrossRef]
79. Mani, S.; Aiyegoro, O.A.; Adeleke, M.A. Association between host genetics of sheep and the rumen microbial composition. *Trop. Anim. Health Prod.* **2022**, *54*, 109. [CrossRef]
80. Lv, W.; Liu, X.; Sha, Y.; Shi, H.; Wei, H.; Luo, Y.; Wang, J.; Li, S.; Hu, J.; Guo, X.; et al. Rumen Fermentation-Microbiota-Host Gene Expression Interactions to Reveal the Adaptability of Tibetan Sheep in Different Periods. *Animals* **2021**, *11*, 3529. [CrossRef] [PubMed]

81. Tanca, A.; Fraumene, C.; Manghina, V.; Palomba, A.; Abbondio, M.; Deligios, M.; Pagnozzi, D.; Addis, M.F.; Uzzau, S. Diversity and functions of the sheep faecal microbiota: A multi-omic characterization. *Microb. Biotechnol.* **2017**, *10*, 541–554. [CrossRef] [PubMed]
82. Wang, L.; Zhang, K.; Zhang, C.; Feng, Y.; Zhang, X.; Wang, X.; Wu, G. Dynamics and stabilization of the rumen microbiome in yearling Tibetan sheep. *Sci. Rep.* **2019**, *9*, 19620. [CrossRef] [PubMed]
83. Asanuma, N.; Yokoyama, S.; Hino, T. Effects of nitrate addition to a diet on fermentation and microbial populations in the rumen of goats, with special reference to *Selenomonas ruminantium* having the ability to reduce nitrate and nitrite. *Anim. Sci. J.* **2015**, *86*, 378–384. [CrossRef]
84. Cremonesi, P.; Conte, G.; Severgnini, M.; Turri, F.; Monni, A.; Capra, E.; Rapetti, L.; Colombini, S.; Chessa, S.; Battelli, G.; et al. Evaluation of the effects of different diets on microbiome diversity and fatty acid composition of rumen liquor in dairy goat. *Animal* **2018**, *12*, 1856–1866. [CrossRef] [PubMed]
85. Fliegerova, K.O.; Podmirseg, S.M.; Vinzelj, J.; Grilli, D.J.; Kvasnova, S.; Schierova, D.; Sechovcova, H.; Mrazek, J.; Siddi, G.; Arenas, G.N.; et al. The Effect of a High-Grain Diet on the Rumen Microbiome of Goats with a Special Focus on Anaerobic Fungi. *Microorganisms* **2021**, *9*, 157. [CrossRef]
86. Peng, X.; Wilken, S.E.; Lankiewicz, T.S.; Gilmore, S.P.; Brown, J.L.; Henske, J.K.; Swift, C.L.; Salamov, A.; Barry, K.; Grigoriev, I.V.; et al. Genomic and functional analyses of fungal and bacterial consortia that enable lignocellulose breakdown in goat gut microbiomes. *Nat. Microbiol.* **2021**, *6*, 499–511. [CrossRef] [PubMed]
87. Min, B.R.; Solaiman, S.; Shange, R.; Eun, J.S. Gastrointestinal Bacterial and Methanogenic Archaea Diversity Dynamics Associated with Condensed Tannin-Containing Pine Bark Diet in Goats Using 16S rDNA Amplicon Pyrosequencing. *Int. J. Microbiol.* **2014**, *2014*, 141909. [CrossRef]
88. Li, B.; Zhang, K.; Li, C.; Wang, X.; Chen, Y.; Yang, Y. Characterization and Comparison of Microbiota in the Gastrointestinal Tracts of the Goat (*Capra hircus*) During Preweaning Development. *Front. Microbiol.* **2019**, *10*, 2125. [CrossRef]
89. Wang, L.; Xu, Q.; Kong, F.; Yang, Y.; Wu, D.; Mishra, S.; Li, Y. Exploring the Goat Rumen Microbiome from Seven Days to Two Years. *PLoS ONE* **2016**, *11*, e0154354. [CrossRef]
90. Zou, X.; Liu, G.; Meng, F.; Hong, L.; Li, Y.; Lian, Z.; Yang, Z.; Luo, C.; Liu, D. Exploring the Rumen and Cecum Microbial Community from Fetus to Adulthood in Goat. *Animals* **2020**, *10*, 1639. [CrossRef] [PubMed]
91. Han, X.; Yang, Y.; Yan, H.; Wang, X.; Qu, L.; Chen, Y. Rumen bacterial diversity of 80 to 110-day-old goats using 16S rRNA sequencing. *PLoS ONE* **2015**, *10*, e0117811. [CrossRef] [PubMed]
92. Zhang, J.; Cai, K.; Mishra, R.; Jha, R. In ovo supplementation of chitooligosaccharide and chlorella polysaccharide affects cecal microbial community, metabolic pathways, and fermentation metabolites in broiler chickens. *Poult. Sci.* **2020**, *99*, 4776–4785. [CrossRef] [PubMed]
93. Palma-Hidalgo, J.M.; Jimenez, E.; Popova, M.; Morgavi, D.P.; Martin-Garcia, A.I.; Yanez-Ruiz, D.R.; Belanche, A. Inoculation with rumen fluid in early life accelerates the rumen microbial development and favours the weaning process in goats. *Anim. Microbiome* **2021**, *3*, 11. [CrossRef]
94. Mammeri, M.; Obregon, D.A.; Chevillot, A.; Polack, B.; Julien, C.; Pollet, T.; Cabezas-Cruz, A.; Adjou, K.T. *Cryptosporidium parvum* Infection Depletes Butyrate Producer Bacteria in Goat Kid Microbiome. *Front. Microbiol.* **2020**, *11*, 548737. [CrossRef]
95. Tong, J.; Ma, W.; Yang, R.; Wang, T.; Chen, X.; Zhang, X.; Tang, X.; Wen, Y.; Chang, J.; Chen, D. Dysbiosis of the gut microbiota maybe exacerbate orf pathology by promoting inflammatory immune responses. *Vet. Microbiol.* **2020**, *251*, 108884. [CrossRef]
96. Wang, L.; Jin, L.; Xue, B.; Wang, Z.; Peng, Q. Characterizing the bacterial community across the gastrointestinal tract of goats: Composition and potential function. *Microbiologyopen* **2019**, *8*, e00820. [CrossRef]
97. Jiang, S.; Huo, D.; You, Z.; Peng, Q.; Ma, C.; Chang, H.; Lin, X.; Wang, L.; Zhang, J. The distal intestinal microbiome of hybrids of Hainan black goats and Saanen goats. *PLoS ONE* **2020**, *15*, e0228496. [CrossRef]
98. Wang, L.; Shah, A.M.; Liu, Y.; Jin, L.; Wang, Z.; Xue, B.; Peng, Q. Relationship between true digestibility of dietary phosphorus and gastrointestinal bacteria of goats. *PLoS ONE* **2020**, *15*, e0225018. [CrossRef]
99. Ren, Z.; Yao, R.; Liu, Q.; Deng, Y.; Shen, L.; Deng, H.; Zuo, Z.; Wang, Y.; Deng, J.; Cui, H.; et al. Effects of antibacterial peptides on rumen fermentation function and rumen microorganisms in goats. *PLoS ONE* **2019**, *14*, e0221815. [CrossRef]
100. Zhuang, Y.; Chai, J.; Cui, K.; Bi, Y.; Diao, Q.; Huang, W.; Usdrowski, H.; Zhang, N. Longitudinal Investigation of the Gut Microbiota in Goat Kids from Birth to Postweaning. *Microorganisms* **2020**, *8*, 1111. [CrossRef] [PubMed]
101. Dou, S.; Gadonna-Widehem, P.; Rome, V.; Hamoudi, D.; Rhazi, L.; Lakhal, L.; Larcher, T.; Bahi-Jaber, N.; Pinon-Quintana, A.; Guyonvarch, A.; et al. Characterisation of Early-Life Fecal Microbiota in Susceptible and Healthy Pigs to Post-Weaning Diarrhoea. *PLoS ONE* **2017**, *12*, e0169851. [CrossRef] [PubMed]
102. Xiao, L.; Estelle, J.; Kiilerich, P.; Ramayo-Caldas, Y.; Xia, Z.; Feng, Q.; Liang, S.; Pedersen, A.O.; Kjeldsen, N.J.; Liu, C.; et al. A reference gene catalogue of the pig gut microbiome. *Nat. Microbiol.* **2016**, *1*, 16161. [CrossRef] [PubMed]
103. Wang, X.; Tsai, T.; Deng, F.; Wei, X.; Chai, J.; Knapp, J.; Apple, J.; Maxwell, C.V.; Lee, J.A.; Li, Y.; et al. Longitudinal investigation of the swine gut microbiome from birth to market reveals stage and growth performance associated bacteria. *Microbiome* **2019**, *7*, 109. [CrossRef]
104. Zhao, G.; Xiang, Y.; Wang, X.; Dai, B.; Zhang, X.; Ma, L.; Yang, H.; Lyu, W. Exploring the Possible Link between the Gut Microbiome and Fat Deposition in Pigs. *Oxid. Med. Cell. Longev.* **2022**, *2022*, 1098892. [CrossRef] [PubMed]

105. Frese, S.A.; Parker, K.; Calvert, C.C.; Mills, D.A. Diet shapes the gut microbiome of pigs during nursing and weaning. *Microbiome* **2015**, *3*, 28. [CrossRef]
106. Klinsoda, J.; Votterl, J.; Koger, S.; Metzler-Zebeli, B.U. Dietary Phytase- and Lactic Acid-Treated Cereals Caused Greater Taxonomic Adaptations than Functional Adaptations in the Cecal Metagenome of Growing Pigs. *Appl. Environ. Microbiol.* **2020**, *87*, e02240-20. [CrossRef]
107. Petry, A.L.; Patience, J.F.; Koester, L.R.; Huntley, N.F.; Bedford, M.R.; Schmitz-Esser, S. Xylanase modulates the microbiota of ileal mucosa and digesta of pigs fed corn-based arabinoxylans likely through both a stimbiotic and prebiotic mechanism. *PLoS ONE* **2021**, *16*, e0246144. [CrossRef]
108. Pollock, J.; Muwonge, A.; Hutchings, M.R.; Mainda, G.; Bronsvoort, B.M.; Gally, D.L.; Corbishley, A. Resistance to change: AMR gene dynamics on a commercial pig farm with high antimicrobial usage. *Sci. Rep.* **2020**, *10*, 1708. [CrossRef]
109. Crespo-Piazuelo, D.; Estelle, J.; Revilla, M.; Criado-Mesas, L.; Ramayo-Caldas, Y.; Ovilo, C.; Fernandez, A.I.; Ballester, M.; Folch, J.M. Characterization of bacterial microbiota compositions along the intestinal tract in pigs and their interactions and functions. *Sci. Rep.* **2018**, *8*, 12727. [CrossRef]
110. De Rodas, B.; Youmans, B.P.; Danzeisen, J.L.; Tran, H.; Johnson, T.J. Microbiome profiling of commercial pigs from farrow to finish. *J. Anim. Sci.* **2018**, *96*, 1778–1794. [CrossRef]
111. Motta, V.; Trevisi, P.; Bertolini, F.; Ribani, A.; Schiavo, G.; Fontanesi, L.; Bosi, P. Exploring gastric bacterial community in young pigs. *PLoS ONE* **2017**, *12*, e0173029. [CrossRef]
112. Verbeek, E.; Keeling, L.; Landberg, R.; Lindberg, J.E.; Dicksved, J. The gut microbiota and microbial metabolites are associated with tail biting in pigs. *Sci. Rep.* **2021**, *11*, 20547. [CrossRef]
113. Borewicz, K.A.; Kim, H.B.; Singer, R.S.; Gebhart, C.J.; Sreevatsan, S.; Johnson, T.; Isaacson, R.E. Changes in the Porcine Intestinal Microbiome in Response to Infection with *Salmonella enterica* and *Lawsonia intracellularis*. *PLoS ONE* **2015**, *10*, e0139106. [CrossRef] [PubMed]
114. Ma, J.; Chen, J.; Gan, M.; Chen, L.; Zhao, Y.; Zhu, Y.; Niu, L.; Zhang, S.; Zhu, L.; Shen, L. Gut Microbiota Composition and Diversity in Different Commercial Swine Breeds in Early and Finishing Growth Stages. *Animals* **2022**, *12*, 1607. [CrossRef] [PubMed]
115. Han, G.G.; Lee, J.Y.; Jin, G.D.; Park, J.; Choi, Y.H.; Chae, B.J.; Kim, E.B.; Choi, Y.J. Evaluating the association between body weight and the intestinal microbiota of weaned piglets via 16S rRNA sequencing. *Appl. Microbiol. Biotechnol.* **2017**, *101*, 5903–5911. [CrossRef] [PubMed]
116. Cremonesi, P.; Biscarini, F.; Castiglioni, B.; Sgoifo, C.A.; Compiani, R.; Moroni, P. Gut microbiome modifications over time when removing in-feed antibiotics from the prophylaxis of post-weaning diarrhea in piglets. *PLoS ONE* **2022**, *17*, e0262199. [CrossRef] [PubMed]
117. Holman, D.B.; Gzyl, K.E.; Mou, K.T.; Allen, H.K. Weaning Age and Its Effect on the Development of the Swine Gut Microbiome and Resistome. *Msystems* **2021**, *6*, e0068221. [CrossRef]
118. Gaio, D.; DeMaere, M.Z.; Anantanawat, K.; Chapman, T.A.; Djordjevic, S.P.; Darling, A.E. Post-weaning shifts in microbiome composition and metabolism revealed by over 25 000 pig gut metagenome-assembled genomes. *Microb. Genom.* **2021**, *7*, 000501. [CrossRef]
119. Guevarra, R.B.; Hong, S.H.; Cho, J.H.; Kim, B.R.; Shin, J.; Lee, J.H.; Kang, B.N.; Kim, Y.H.; Wattanaphansak, S.; Isaacson, R.E.; et al. The dynamics of the piglet gut microbiome during the weaning transition in association with health and nutrition. *J Anim. Sci. Biotechnol.* **2018**, *9*, 54. [CrossRef] [PubMed]
120. Fenske, G.J.; Ghimire, S.; Antony, L.; Christopher Hennings, J.; Scaria, J. Integration of culture-dependent and independent methods provides a more coherent picture of the pig gut microbiome. *FEMS Microbiol. Ecol.* **2020**, *96*, fiaa022. [CrossRef]
121. Munk, P.; Andersen, V.D.; de Knegt, L.; Jensen, M.S.; Knudsen, B.E.; Lukjancenko, O.; Mordhorst, H.; Clasen, J.; Agerso, Y.; Folkesson, A.; et al. A sampling and metagenomic sequencing-based methodology for monitoring antimicrobial resistance in swine herds. *J. Antimicrob. Chemother.* **2017**, *72*, 385–392. [CrossRef] [PubMed]
122. Gweon, H.S.; Shaw, L.P.; Swann, J.; De Maio, N.; AbuOun, M.; Niehus, R.; Hubbard, A.T.M.; Bowes, M.J.; Bailey, M.J.; Peto, T.E.A.; et al. The impact of sequencing depth on the inferred taxonomic composition and AMR gene content of metagenomic samples. *Environ. Microbiome* **2019**, *14*, 7. [CrossRef] [PubMed]
123. Wylezich, C.; Belka, A.; Hanke, D.; Beer, M.; Blome, S.; Hoper, D. Metagenomics for broad and improved parasite detection: A proof-of-concept study using swine faecal samples. *Int. J. Parasitol.* **2019**, *49*, 769–777. [CrossRef]
124. Ramayo-Caldas, Y.; Prenafeta-Boldu, F.; Zingaretti, L.M.; Gonzalez-Rodriguez, O.; Dalmau, A.; Quintanilla, R.; Ballester, M. Gut eukaryotic communities in pigs: Diversity, composition and host genetics contribution. *Anim. Microbiome* **2020**, *2*, 18. [CrossRef] [PubMed]
125. Mann, E.; Schmitz-Esser, S.; Zebeli, Q.; Wagner, M.; Ritzmann, M.; Metzler-Zebeli, B.U. Mucosa-associated bacterial microbiome of the gastrointestinal tract of weaned pigs and dynamics linked to dietary calcium-phosphorus. *PLoS ONE* **2014**, *9*, e86950. [CrossRef] [PubMed]
126. Tang, S.; Xin, Y.; Ma, Y.; Xu, X.; Zhao, S.; Cao, J. Screening of Microbes Associated With Swine Growth and Fat Deposition Traits Across the Intestinal Tract. *Front. Microbiol.* **2020**, *11*, 586776. [CrossRef] [PubMed]
127. Wang, W.; Hu, H.; Zijlstra, R.T.; Zheng, J.; Gänzle, M.G. Metagenomic reconstructions of gut microbial metabolism in weanling pigs. *Microbiome* **2019**, *7*, 48. [CrossRef]

128. Wang, C.; Wei, S.; Chen, N.; Xiang, Y.; Wang, Y.; Jin, M. Characteristics of gut microbiota in pigs with different breeds, growth periods and genders. *Microb. Biotechnol.* **2022**, *15*, 793–804. [CrossRef]
129. Wylensek, D.; Hitch, T.; Riedel, T.; Afrizal, A.; Kumar, N.; Wortmann, E.; Liu, T.; Devendran, S.; Lesker, T.R.; Hernández, S.B.; et al. A collection of bacterial isolates from the pig intestine reveals functional and taxonomic diversity. *Nat. Commun.* **2020**, *11*, 6389. [CrossRef]
130. Qu, A.; Brulc, J.M.; Wilson, M.K.; Law, B.F.; Theoret, J.R.; Joens, L.A.; Konkel, M.E.; Angly, F.; Dinsdale, E.A.; Edwards, R.A.; et al. Comparative metagenomics reveals host specific metavirulomes and horizontal gene transfer elements in the chicken cecum microbiome. *PLoS ONE* **2008**, *3*, e2945. [CrossRef] [PubMed]
131. Danzeisen, J.L.; Kim, H.B.; Isaacson, R.E.; Tu, Z.J.; Johnson, T.J. Modulations of the chicken cecal microbiome and metagenome in response to anticoccidial and growth promoter treatment. *PLoS ONE* **2011**, *6*, e27949. [CrossRef] [PubMed]
132. Choi, K.Y.; Lee, T.K.; Sul, W.J. Metagenomic Analysis of Chicken Gut Microbiota for Improving Metabolism and Health of Chickens—A Review. *Asian-Australas. J. Anim. Sci.* **2015**, *28*, 1217–1225. [CrossRef]
133. Durazzi, F.; Sala, C.; Castellani, G.; Manfreda, G.; Remondini, D.; De Cesare, A. Comparison between 16S rRNA and shotgun sequencing data for the taxonomic characterization of the gut microbiota. *Sci. Rep.* **2021**, *11*, 3030. [CrossRef]
134. De Cesare, A.; Sirri, F.; Manfreda, G.; Moniaci, P.; Giardini, A.; Zampiga, M.; Meluzzi, A. Effect of dietary supplementation with *Lactobacillus acidophilus* D2/CSL (CECT 4529) on caecum microbioma and productive performance in broiler chickens. *PLoS ONE* **2017**, *12*, e0176309. [CrossRef]
135. Gilroy, R.; Ravi, A.; Getino, M.; Pursley, I.; Horton, D.L.; Alikhan, N.F.; Baker, D.; Gharbi, K.; Hall, N.; Watson, M.; et al. Extensive microbial diversity within the chicken gut microbiome revealed by metagenomics and culture. *PeerJ* **2021**, *9*, e10941. [CrossRef] [PubMed]
136. Medvecky, M.; Cejkova, D.; Polansky, O.; Karasova, D.; Kubasova, T.; Cizek, A.; Rychlik, I. Whole genome sequencing and function prediction of 133 gut anaerobes isolated from chicken caecum in pure cultures. *BMC Genom.* **2018**, *19*, 561. [CrossRef] [PubMed]
137. Xu, P.; Shi, Y.; Liu, P.; Yang, Y.; Zhou, C.; Li, G.; Luo, J.; Zhang, C.; Cao, H.; Hu, G.; et al. 16S rRNA gene sequencing reveals an altered composition of the gut microbiota in chickens infected with a nephropathogenic infectious bronchitis virus. *Sci. Rep.* **2020**, *10*, 3556. [CrossRef]
138. Clavijo, V.; Florez, M.J.V. The gastrointestinal microbiome and its association with the control of pathogens in broiler chicken production: A review. *Poult. Sci.* **2018**, *97*, 1006–1021. [CrossRef]
139. Such, N.; Farkas, V.; Csitari, G.; Pal, L.; Marton, A.; Menyhart, L.; Dublecz, K. Relative Effects of Dietary Administration of a Competitive Exclusion Culture and a Synbiotic Product, Age and Sampling Site on Intestinal Microbiota Maturation in Broiler Chickens. *Vet. Sci.* **2021**, *8*, 187. [CrossRef]
140. Adenaike, A.S.; Akpan, U.; Awopejo, O.O.; Oloye, O.S.; Alli-Balogun, A.O.; Agbaje, M.; Ikeobi, C.O.N. Characterization of the cecal microbiome composition of Nigerian indigenous chickens. *Trop. Anim. Health Prod.* **2022**, *54*, 211. [CrossRef] [PubMed]
141. Lee, S.J.; Cho, S.; La, T.M.; Lee, H.J.; Lee, J.B.; Park, S.Y.; Song, C.S.; Choi, I.S.; Lee, S.W. Comparison of microbiota in the cloaca, colon, and magnum of layer chicken. *PLoS ONE* **2020**, *15*, e0237108. [CrossRef] [PubMed]
142. Chen, S.; Xiang, H.; Zhang, H.; Zhu, X.; Wang, D.; Wang, J.; Yin, T.; Liu, L.; Kong, M.; Li, H.; et al. Rearing system causes changes of behavior, microbiome, and gene expression of chickens. *Poult. Sci.* **2019**, *98*, 3365–3376. [CrossRef] [PubMed]
143. Sun, J.; Liao, X.P.; D'Souza, A.W.; Boolchandani, M.; Li, S.H.; Cheng, K.; Luis Martinez, J.; Li, L.; Feng, Y.J.; Fang, L.X.; et al. Environmental remodeling of human gut microbiota and antibiotic resistome in livestock farms. *Nat. Commun.* **2020**, *11*, 1427. [CrossRef]
144. Ramírez, G.A.; Richardson, E.; Clark, J.; Keshri, J.; Drechsler, Y.; Berrang, M.E.; Meinersmann, R.J.; Cox, N.A.; Oakley, B.B. Broiler chickens and early life programming: Microbiome transplant-induced cecal community dynamics and phenotypic effects. *PLoS ONE* **2020**, *15*, e0242108. [CrossRef]
145. Kabir, M.H.B.; Han, Y.; Lee, S.H.; Nugraha, A.B.; Recuenco, F.; Murakoshi, F.; Xuan, X.; Kato, K. Prevalence and molecular characterization of *Cryptosporidium* species in poultry in Bangladesh. *One Health* **2020**, *9*, 100122. [CrossRef]
146. Wei, S.; Morrison, M.; Yu, Z. Bacterial census of poultry intestinal microbiome. *Poult. Sci.* **2013**, *92*, 671–683. [CrossRef]
147. Saxena, S.; Saxena, V.; Tomar, S.; Sapcota, D.; Gonmei, G. Characterisation of caecum and crop microbiota of Indian indigenous chicken targeting multiple hypervariable regions within 16S rRNA gene. *Br. Poult. Sci.* **2016**, *57*, 381–389. [CrossRef]
148. Biasato, I.; Ferrocino, I.; Grego, E.; Dabbou, S.; Gai, F.; Gasco, L.; Cocolin, L.; Capucchio, M.T.; Schiavone, A. Gut Microbiota and Mucin Composition in Female Broiler Chickens Fed Diets including Yellow Mealworm (*Tenebrio molitor*, L.). *Animals* **2019**, *9*, 213. [CrossRef] [PubMed]
149. Yang, C.; Diarra, M.S.; Choi, J.; Rodas-Gonzalez, A.; Lepp, D.; Liu, S.; Lu, P.; Mogire, M.; Gong, J.; Wang, Q.; et al. Effects of encapsulated cinnamaldehyde on growth performance, intestinal digestive and absorptive functions, meat quality and gut microbiota in broiler chickens. *Transl. Anim. Sci.* **2021**, *5*, txab099. [CrossRef]
150. Videnska, P.; Faldynova, M.; Juricova, H.; Babak, V.; Sisak, F.; Havlickova, H.; Rychlik, I. Chicken faecal microbiota and disturbances induced by single or repeated therapy with tetracycline and streptomycin. *BMC Vet. Res.* **2013**, *9*, 30. [CrossRef] [PubMed]
151. D'Costa, V.M.; McGrann, K.M.; Hughes, D.W.; Wright, G.D. Sampling the antibiotic resistome. *Science* **2006**, *311*, 374–377. [CrossRef] [PubMed]

152. Skandalis, N.; Maeusli, M.; Papafotis, D.; Miller, S.; Lee, B.; Theologidis, I.; Luna, B. Environmental Spread of Antibiotic Resistance. *Antibiotics* **2021**, *10*, 640. [CrossRef]
153. Kim, D.W.; Cha, C.J. Antibiotic resistome from the One-Health perspective: Understanding and controlling antimicrobial resistance transmission. *Exp. Mol. Med.* **2021**, *53*, 301–309. [CrossRef]
154. Liu, J.; Taft, D.H.; Maldonado-Gomez, M.X.; Johnson, D.; Treiber, M.L.; Lemay, D.G.; DePeters, E.J.; Mills, D.A. The fecal resistome of dairy cattle is associated with diet during nursing. *Nat. Commun.* **2019**, *10*, 4406. [CrossRef]
155. Lopez-Catalina, A.; Atxaerandio, R.; Garcia-Rodriguez, A.; Goiri, I.; Gutierrez-Rivas, M.; Jimenez-Montero, J.A.; Gonzalez-Recio, O. Characterisation of the rumen resistome in Spanish dairy cattle. *Anim. Microbiome* **2021**, *3*, 63. [CrossRef]
156. Ma, T.; McAllister, T.A.; Guan, L.L. A review of the resistome within the digestive tract of livestock. *J. Anim. Sci. Biotechnol.* **2021**, *12*, 121. [CrossRef]
157. Hitch, T.C.A.; Thomas, B.J.; Friedersdorff, J.C.A.; Ougham, H.; Creevey, C.J. Deep sequence analysis reveals the ovine rumen as a reservoir of antibiotic resistance genes. *Environ. Pollut.* **2018**, *235*, 571–575. [CrossRef]
158. Wang, Y.; Hu, Y.; Liu, F.; Cao, J.; Lv, N.; Zhu, B.; Zhang, G.; Gao, G.F. Integrated metagenomic and metatranscriptomic profiling reveals differentially expressed resistomes in human, chicken, and pig gut microbiomes. *Environ. Int.* **2020**, *138*, 105649. [CrossRef]
159. Koorakula, R.; Schiavinato, M.; Ghanbari, M.; Wegl, G.; Grabner, N.; Koestelbauer, A.; Klose, V.; Dohm, J.C.; Domig, K.J. Metatranscriptomic Analysis of the Chicken Gut Resistome Response to In-Feed Antibiotics and Natural Feed Additives. *Front. Microbiol.* **2022**, *13*, 833790. [CrossRef] [PubMed]
160. Murray, M.; Salvatierra, G.; Davila-Barclay, A.; Ayzanoa, B.; Castillo-Vilcahuaman, C.; Huang, M.; Pajuelo, M.J.; Lescano, A.G.; Cabrera, L.; Calderon, M.; et al. Market Chickens as a Source of Antibiotic-Resistant *Escherichia coli* in a Peri-Urban Community in Lima, Peru. *Front. Microbiol.* **2021**, *12*, 635871. [CrossRef] [PubMed]
161. Langille, M.G.I.; Zaneveld, J.; Caporaso, J.G.; McDonald, D.; Knights, D.; Reyes, J.A.; Clemente, J.C.; Burkepile, D.E.; Vega Thurber, R.L.; Knight, R.; et al. Predictive functional profiling of microbial communities using 16S rRNA marker gene sequences. *Nat. Biotech.* **2013**, *31*, 814–821. [CrossRef] [PubMed]
162. Douglas, G.M.; Maffei, V.J.; Zaneveld, J.; Yurgel, S.N.; Brown, J.R.; Taylor, C.M.; Huttenhower, C.; Langille, M.G. PICRUSt2: An improved and extensible approach for metagenome inference. *BioRxiv* **2019**, 672295. [CrossRef]
163. Louca, S.; Parfrey, L.W.; Doebeli, M. Decoupling function and taxonomy in the global ocean microbiome. *Science* **2016**, *353*, 1272–1277. [CrossRef]
164. Wemheuer, F.; Taylor, J.A.; Daniel, R.; Johnston, E.; Meinicke, P.; Thomas, T.; Wemheuer, B. Tax4Fun2: A R-based tool for the rapid prediction of habitat-specific functional profiles and functional redundancy based on 16S rRNA gene marker gene sequences. *Environ. Microbiome* **2020**, *15*, 11. [CrossRef] [PubMed]
165. Sun, S.; Jones, R.B.; Fodor, A.A. Inference-based accuracy of metagenome prediction tools varies across sample types and functional categories. *Microbiome* **2020**, *8*, 46. [CrossRef] [PubMed]
166. Toole, D.R.; Zhao, J.; Martens-Habbena, W.; Strauss, S.L. Bacterial functional prediction tools detect but underestimate metabolic diversity compared to shotgun metagenomics in southwest Florida soils. *Appl. Soil Ecol.* **2021**, *168*, 104129. [CrossRef]
167. Sun, B.; Hou, L.; Yang, Y. The Development of the Gut Microbiota and Short Chain Fatty Acids of Layer Chickens in Different Growth Periods. *Front. Vet. Sci.* **2021**, *8*, 666535. [CrossRef]
168. Kogut, M.H.; Arsenault, R.J. Editorial: Gut Health: The New Paradigm in Food Animal Production. *Front. Vet. Sci.* **2016**, *3*, 71. [CrossRef]
169. Kraimi, N.; Dawkins, M.; Gebhardt-Henrich, S.G.; Velge, P.; Rychlik, I.; Volf, J.; Creach, P.; Smith, A.; Colles, F.; Leterrier, C. Influence of the microbiota-gut-brain axis on behavior and welfare in farm animals: A review. *Physiol. Behav.* **2019**, *210*, 112658. [CrossRef]
170. O'Hara, E.; Neves, A.; Song, Y.; Guan, L.L. The Role of the Gut Microbiome in Cattle Production and Health: Driver or Passenger? *Annu. Rev. Anim. Biosci.* **2020**, *8*, 199–220. [CrossRef] [PubMed]
171. Diaz, J.; Reese, A.T. Possibilities and limits for using the gut microbiome to improve captive animal health. *Anim. Microbiome* **2021**, *3*, 89. [CrossRef] [PubMed]
172. Chen, S.; Luo, S.; Yan, C. Gut Microbiota Implications for Health and Welfare in Farm Animals: A Review. *Animals* **2022**, *12*, 93. [CrossRef] [PubMed]
173. Chen, B.; Li, D.; Leng, D.; Kui, H.; Bai, X.; Wang, T. Gut microbiota and meat quality. *Front. Microbiol.* **2022**, *13*, 951726. [CrossRef] [PubMed]
174. Lourenco, J.M.; Welch, C.B. Using microbiome information to understand and improve animal performance. *Ital. J. Anim. Sci.* **2022**, *21*, 899–913. [CrossRef]
175. Khalil, A.; Batool, A.; Arif, S. Healthy Cattle Microbiome and Dysbiosis in Diseased Phenotypes. *Ruminants* **2022**, *2*, 134–156. [CrossRef]
176. Clemmons, B.A.; Voy, B.H.; Myer, P.R. Altering the Gut Microbiome of Cattle: Considerations of Host-Microbiome Interactions for Persistent Microbiome Manipulation. *Microb. Ecol.* **2019**, *77*, 523–536. [CrossRef]
177. Malmuthuge, N.; Guan, L.L. Understanding the gut microbiome of dairy calves: Opportunities to improve early-life gut health. *J. Dairy Sci.* **2017**, *100*, 5996–6005. [CrossRef]
178. Campanaro, S.; Treu, L.; Kougias, P.G.; Zhu, X.; Angelidaki, I. Taxonomy of anaerobic digestion microbiome reveals biases associated with the applied high throughput sequencing strategies. *Sci. Rep.* **2018**, *8*, 1926. [CrossRef]

MDPI AG
Grosspeteranlage 5
4052 Basel
Switzerland
Tel.: +41 61 683 77 34

Animals Editorial Office
E-mail: animals@mdpi.com
www.mdpi.com/journal/animals